AMERICAN ENVIRONMENTAL FICTION, 1782–1847

To Alisa, Allston, and Freya

American Environmental Fiction, 1782–1847

MATTHEW WYNN SIVILS
Iowa State University, USA

Routledge
Taylor & Francis Group
LONDON AND NEW YORK

First published in paperback 2024

First published 2014 by Ashgate Publishing

Published 2016 by Routledge
4 Park Square, Milton Park, Abingdon, Oxon OX14 4RN

and by Routledge
605 Third Avenue, New York, NY 10158

Routledge is an imprint of the Taylor & Francis Group, an informa business

Publisher's Note
The publisher has gone to great lengths to ensure the quality of this reprint but points out that some imperfections in the original copies may be apparent.

British Library Cataloguing in Publication Data
A catalogue record for this book is available from the British Library

The Library of Congress has cataloged the printed edition as follows:
Sivils, Matthew Wynn, 1971–
American Environmental Fiction, 1782–1847 / by Matthew Wynn Sivils.
pages cm
Includes bibliographical references and index.
ISBN 978-1-4094-3163-3 (hardcover: alk. paper)
1. American literature—18th century—History and criticism. 2. American literature—19th century—History and criticism. 3. Ecofiction, American—History and criticism. 4. Environmentalism in literature. 5. Ecology in literature. I. Title.
PS169.E25S55 2015
810.9'36—dc22
2014006147

ISBN: 978-1-4094-3163-3 (hbk)
ISBN: 978-1-03-292812-8 (pbk)
ISBN: 978-1-315-56667-2 (ebk)

DOI: 10.4324/9781315566672

Contents

List of Figures

Acknowledgments

For their advice at varying stages of this project, I thank Allan Axelrad, Wayne Franklin, John McWilliams, Lance Schachterle, Jeffrey Walker, and Laura E. Wasowicz. I am also indebted to Ann Donahue, Seth F. Hibbert, and the rest of the Ashgate Publishing staff.

I thank Iowa State University for their generous support, which included a Gregory L. and Kathleen C. Geoffroy Faculty Fellowship, a Center for Excellence in the Arts and Humanities Fellowship, a College of Liberal Arts and Sciences Small Grant, a Publication Subvention Grant, and a Faculty Professional Development Assignment. I also owe a debt of gratitude to the American Antiquarian Society, which awarded me a Justin G. Schiller Fellowship as well as a stint as scholar-in-residence. For permitting me to reprint images from their collections, I thank the American Antiquarian Society, the British Museum of Natural History, the Metropolitan Museum of Art, and the Corcoran Gallery of Art.

Two chapters were originally published elsewhere in slightly different form. A portion of Chapter 1 first appeared as "William Bartram's *Travels* and the Rhetoric of Ecological Communities" in *Interdisciplinary Studies in Literature and Environment* 11.1 (2004): 56–70. Chapter 5 was originally published as "Doctor Bat's Ass: Buffon, American Degeneracy, and Cooper's *The Prairie*" in *Western American Literature* 44.4 (2010): 343–61. I thank these journals for allowing me to reprint this material.

I reserve my deepest thanks for my wife, Alisa, and our children, Allston and Freya.

Introduction

> [W]hat could they see but a hideous and desolate wilderness, full of wild beasts and wild men ... the whole country, full of woods and thickets, represented a wild and savage hue. If they looked behind them, there was the mighty ocean which they had passed and was now a main bar and gulf to separate them from all the civil parts of the world.
>
> —William Bradford, from *Of Plymouth Plantation* (1630–1651)[1]

> The passage of the Patowmac through the Blue ridge is perhaps one of the most stupendous scenes in nature. This scene is worth a voyage across the Atlantic.
>
> —Thomas Jefferson, from *Notes on the State of Virginia* (1782–1787)[2]

As these epigraphs suggest, the European American literary response to the land had shifted dramatically by the time Jefferson and company set into motion the political machinery of the fledgling United States. Waves had long erased the first boot prints from the shores of what European colonizers had foolishly called the New World. By then the marks on the land had taken a much more permanent form. In what has since become something of a characteristically American pathology, as writers like Jefferson glorified the land's beauty, the degradation of that land—along with the extermination of its indigenous population—continued apace. Forests and buffalo fell to the percussive chorus of axe and rifle. Rivers clotted with mills and towns.[3] The land that served as muse to the nation's cultural formation became the very fabric and fuel of its identity. Still, the natural world (or its nostalgic portrayal) was increasingly depicted in the pages of a growing national literature, one created by writers desperate to forge a new culture for a new nation, and fed by the destruction of that which made it most distinct from its European roots.

In the pages that follow, I look at how some of these writers, specifically those creating the nation's first works of fiction, reconstituted, reshaped, and reimagined a natural world that seemed to change faster than their ink could dry. At its core this study seeks to answer this question: How did the fiction and proto-fictional texts of the early United States influence, as well as reflect, the nation's new environmental consciousness? I seek answers to this question by studying a

---

1 Bradford, *Of Plymouth Plantation: 1620–1647*, 62.

2 Jefferson, *Notes on the State of Virginia*, 142–3.

3 Here I invoke in passing the concept of the human experience of working the land, and while I do not focus on this issue in my study, Timothy Sweet does in his *American Georgics: Economy and Environment in Early American Literature*, in which he studies a handful of early American texts dealing with the Georgic theme of nature as a "site of labor" (that is, farming, wood-cutting, and market hunting) rather than the more commonly studied pastoral that presents the natural world as a place of leisure (2).

modest selection of early republic fictional or pseudo-fictional narratives in which the non-human environment plays more than a passing role.[4] The environmental literature of the early United States is a rich cache of cultural artifacts, one that betters our understanding of the origins and development of the nation's larger environmental mind. It does so while also telling us as much, if not more, about the myth of America as imagined by a handful of writers bent on conjuring a new culture from what they saw in the land. Writers of the early national period were understandably concerned with developing a unique idea of just what constituted the European American experience within its associated and embryonic national identity. The striking landscapes of North America left writers with no real alternative but to take them as their subject, to cast them as one of the defining elements of what European American writers insisted was their world, despite the copious evidence that it had been (and in many areas still was) thoroughly inhabited by Native American communities.

As a traveler crossed the land from east to west the landscape transitioned from coastal towns and settlements to farms that gave way to wilderness, and then to the mysterious openness of the plains, and, even further, to imagined landscapes of abstract promise and romantic mystery. It was an assemblage of environments at times radically different from those of Europe, and as the tendrils of white settlement entwined the land, they strangled Native American cultures and turned the natural resources to their imported purposes. In part, it became the job of this new literature of the land to incorporate, legitimize, or otherwise obscure that conquest. "There was," as David Mazel writes,

> a particularly urgent need for such a new "American" subjectivity during the early national period, when it was widely felt that the discovery of a "natural" or "latent" American identity might justify the rebellion against Britain and legitimate the nation created afterward ... It was not only the new nation's literature, however, that was pressed into the service of the national narrative; its very *environment* was called on as well.[5]

Thus it was perhaps inevitable that American writers would, as they strove to create a culture ostensibly independent of its European origins, pay special attention to the nature of their newly acquired territory and to ascribe to it a key role in that project of national validation. This project of leveraging the imagery of the natural

---

4 As my study focuses solely on the portrayal of the environment in fictional prose from 1782 to 1847, I do not intend my selection of texts to represent the whole of environmental literature of the period. One book that is particularly helpful in completing the picture of the environmental culture of the period is Karen Kilcup's *Fallen Forests: Emotion, Embodiment, and Ethics in American Women's Environmental Writing, 1781–1924*. Kilcup's examination of underappreciated women nature writers and poets is particularly strong in its analysis of the nature writings (or environmental elements found in the works of) of Mary Jemison, Harriet Jacobs, and Caroline Kirkland to name just a few.

5 Mazel, *American Literary Environmentalism*, xviii–xix.

world to give shape to the national identity of the United States was probably not entirely intentional or formalized, but the cultural void needed to be filled, and America's bookshelves were a good place to start.

These early texts are, however, a problematic body of work, especially because they occupy a proto-environmental moment in the American literary tradition. Environmental literature of this period simply has not enjoyed the same critical attention as works dating from the second half of the nineteenth century. As Michael P. Branch has pointed out, early American texts are too often overlooked because many ecocritics needlessly adhere to a: "circumscribed generic definition of nature writing" that results in a situation in which "most critical studies of American environmental literature examine works written since the mid-nineteenth century, often citing Henry David Thoreau as the progenitor of the American nature writing tradition."[6] Branch makes a good argument for why we need to expand our definition of nature writing to include the galaxy of documents that preceded Thoreau's work and that also extend beyond merely the personal essay. Branch, who is mostly concerned with pre-1800 works, includes in his list such understudied but environmentally promising texts as "the relation, report, sermon, tract, letter, providential history, spiritual autobiography, or personal diary."[7] These are indeed rich areas for ecocritical analysis, but for the purposes of this study I focus on the fiction that arose during the early republic period. It is a body of work that surfaced during a key moment in American literary and environmental history, one in which authors increasingly concerned themselves with linking the dawning nation's natural world to its bloody past and uncertain future.

One problem we run into when studying these texts is the fact that the very terms we use to examine them are far from ideal. With this challenge in mind, it is helpful to look specifically at two of the terms found in the title of this book: *American* and *environmental*. To begin, I am well aware of how recent hemispheric and transnational approaches to studying the Anglophone literary tradition productively complicate the use of a term like *American* as a cultural signifier. While I admire the questioning of such terms to better our understanding of the fluid nature of national boundaries and identities, for this study, one that deals solely with texts written in and about the environment of the United States, I adhere to the conventional definition of "American" to indicate a text, figure, or concept originating from the political entity of the United States.

In a similar vein, like other twenty-first-century literary scholars, I employ the present-day term *environmental* to categorize these works, which variously take as their subjects natural spaces and their non-human elements. It is, however, crucial to remember that, with the possible exception of William Bartram, none of the

[6] Branch, "Before Nature Writing: Discourses of Colonial American Natural History," 91.

[7] Branch, "Before Nature Writing: Discourses of Colonial American Natural History," 92.

writers I examine in this book would have identified themselves as particularly *environmental.* They would not even recognize our present-day use of the term, which simply was not in their collective vocabulary. A search of digital copies reveals that, as amazing as it may seem, the words *environment* and *environmental* never appear in any of the texts that form the subject of this book—not in William Bartram's *Travels* (1791), not in J. Hector St. John de Crèvecoeur's *Letters from an American Farmer* (1782), nor even in Charles Brockden Brown's *Edgar Huntly* (1799) or any of the James Fenimore Cooper novels I examine. Instead, these writers called it *the land, the landscape, the wilderness*, or, in its broadest sense—*nature*.

While I, like other literary scholars, use the term *environment* as a genre designation and as an occasional stand-in for these other terms, I am also cognizant of the fact that there is more than just a semantic difference between the twenty-first-century definition of *environment* and the nineteenth-century conception of *nature*. The meanings of these terms also differ substantially even today, depending upon how they are interpreted. For this study, however, I define as *environmental* that portion of the non-human world that serves as a physical and aesthetic nexus. In this sense, the environment is a site of transference, where objects of nature interact with human subjects (and vice versa). This interaction results in a modification of both the human and the land, with neither quite the same after coming into contact. My view is supported by David Mazel, who, in *American Literary Environmentalism*, offers an illuminating etymology and reconsideration of the term, one that asks how *environment* reveals a range of underlying critical biases. He writes, in part, "we need to stop asking what the environment *is*, stop trying to enumerate its elements, and ask instead, How is it that these enumerated elements have come to environ *us*?"[8] Mazel continues by asking us to reconsider the way the term *environment* situates and labels the world as experienced by the human subject. He writes,

> The answer lies in the *presumption of the presence of*, and ultimately in the *act of entry by*, the speakers who can sensibly say, "*our* environment." It is not any action on the part of our surroundings that has made them *our* surroundings, but the onset of and the continuation of our being here: a matter not of ontology but of history.[9]

Thus my use of the term *environment* in this book denotes a specific location of impact between the nonhuman and human spheres.

This definition of *environment* demonstrates how works of fiction differed substantially between the first days of the Early Republic and the publication of Ralph Waldo Emerson's *Nature*. Recognizing the slippery character of the early nineteenth-century definition of *nature*, Emerson opens his famous essay with his own take on the term. For Emerson, *nature* bears both a "philosophical" and

[8] Mazel, *American Literary Environmentalism*, 36–7.

[9] Mazel, *American Literary Environmentalism*, 37.

a "common" meaning. For him the term philosophically designates: "all that is separate from us, all which Philosophy distinguishes as the NOT ME, that is, both nature and art, all other men and my own body" while "in the common sense, [it] refers to essences unchanged by man; space, the air, the river, the leaf."[10] Thus Emerson's definition of *Nature/nature* is essentially the opposite of my definition of *environment*. It is notable that in our current age of widespread pollution, Emerson's common definition of nature is no longer valid. Today one can find no "essences," no river, no leaf left unchanged or untainted by humanity. Yet even in Emerson's day there was already a growing concern about how European colonization and poor agricultural practices were degrading the environment. Citing the travel writing of Jeremy Belknap, Timothy Dwight, and Marquis de Chastellux, Lloyd Willis writes that "By the dawning of the nineteenth century, American and European observers were certain that the North American environment was experiencing major changes—even climactic changes—due to European involvement with the continent."[11] Hence, one of the more pronounced weaknesses inherent to Emerson's definition of Nature is the idea that the natural world is somehow a place devoid of human influence or transference. In Emerson's outdated terms, Nature is the realm of the "NOT ME," but as is clear by the complexities of our present-day environmental concerns, humanity can no longer afford the delusion that we are not intimately and inextricably connected to the land. As I will repeatedly argue in this book, this decidedly non-Emersonian, and far more environmentally practical, understanding of how the nonhuman world influences (and is itself changed by) human practices emerges in several works of early American environmental literature.

In fact, academics seem to increasingly realize that terms like *nature* may no longer mean much, if anything, at all, and not because humans have destroyed nature but because in coming to appreciate a more mature ecological worldview we accept that both the Emersonian "NOT ME" and the "me" are inextricably connected parts of the ecological whole. The ambiguous catch-all term of *nature* might not only be obsolete but also a detriment, in that it could hinder more productive ways of thinking ecologically. In his thought-provoking book *Ecology without Nature*, Timothy Morton pulls no punches when he argues that "'nature' is an arbitrary rhetorical construct, empty of independent, genuine existence behind or beyond the texts we create about it." Morton makes a similar argument for much of the terminology and practice of ecocriticism, including the use of the term *environment*.[12] While I do not go as far as Morton in shelving these familiar concepts, I do think it productive to question why and how we employ these terms. For the purposes of this study, I find I must label texts in a way that makes sense to present-day readers, and so in discussing these writers I follow the critical herd

10 Emerson, *Nature*, 36.

11 Willis, *Environmental Evasion*, 6.

12 Morton, *Ecology without Nature*, 21–2. See also Fletcher, *A New Theory for American Poetry*, 122–3.

and adopt a term that those early writers would have found almost entirely alien. I take solace in the idea that such linguistic compromises are unavoidable and probably occur in most cases in which scholars study early texts.[13] It is not ideal, but as long as we remain aware of the problems inherent in the terminology, the damage should be minimal.

Another challenge we face when looking at such early American environmental texts is that few if any are what we might think of as examples of Leopoldian ecocentricism, in which the species *Homo sapiens* represents simply a piece of the ecological puzzle, and the components of the natural world are no longer merely resources but are equal beneficiaries of our concept of morality. As our cavalcade of present-day environmental crises attests, too few people even now can conceive of, much less adhere to, this kind of Leopoldian land ethic. To ask any of the environmental writers examined in this book to promote such a mature and complex ecological accountability would be to engage in a form of presentism bordering on folly. After all, each wrote before the publication of Charles Darwin's *Origin of Species* and well before the creation of modern ecological thought. Such presentism is one of the few weaknesses found in Lawrence Buell's oft-cited ingredients for an environmental text, the third of which is "*Human accountability to the environment is part of the text's ethical orientation*."[14] This kind of accountability is predictably absent from many of the works found in this study simply because most of these writers saw before them an environment of seemingly inexhaustible plenty. That humankind could ever hunt a species to extinction, ever degrade an entire environment, ever subvert large-scale natural processes was to most of them inconceivable, especially before the publication of James Fenimore Cooper's environmental landmark *The Pioneers* (1823), a novel that, in directly calling for such accountability, reshaped the possibilities of the environmental literary genre.

Most ecocritics have since abandoned or muted their reliance upon Buell's original definition. Buell himself, about ten years after the publication of his list, recognized how it had contributed to a needlessly limited conception of what constitutes an environmental text. He writes,

> Once I thought it helpful to try to specify a subspecies of "environmental text" ... Now, it seems to me more productive to think inclusively of environmentality as a property of any text—to maintain that all human artifacts bear such traces, and at several stages: in the composition, the embodiment, and the reception.[15]

---

13 In her recent study of early American women's nature writing and poetry, Kilcup offers a particularly thought-provoking discussion of the problems inherent to terms such as "nature" and "environmental" when considered from a multiethnic and gendered point of view, see *Fallen Forests*, 6–8.

14 Buell, *The Environmental Imagination*, 7.

15 Buell, *The Future of Environmental Criticism*, 25.

Buell's newer definition is notable not only for its inclusivity but also for its recognition that texts display environmentality both in their content and also in the realities of their production and physical states. While this study is concerned predominantly with the textual content of early American environmental fiction, Buell's recognition that texts can be classified as more or less environmental based on the stages of production is a promising avenue of inquiry, especially as it melds with recent gains in scholarly attention to the history of the book as object.

Central to this study, however, is the recognition that depending upon how they are read, any number of literary texts might be said to contain significant naturalistic characteristics that justify their classification as environmental, even in cases where the text falls far outside of Buell's original attempt at a definition. It is tempting, then, to envision an imaginary gradient of literary environmentality, one upon which all texts might be ranked from less to more environmental. However, such a ranking, even if it were possible, would be of little actual use, because it would mirror the very problems found in a more prescriptive and presentist set of criteria. In attempting to order texts based upon a subjective and ever-shifting evaluation of their environmentality we would risk overlooking subtly important environmental aspects in those texts ranked at the bottom of the scale. For example, on such a scale Charles Brockden Brown's *Edgar Huntly*, with its labyrinthine forest scenes, cavern adventures, and panther attacks, would probably rank above Brown's *Arthur Mervyn* (1798–1800), but such a categorization would overlook the great value of investigating the environmental facets woven into Brown's reimagining of the metaphorical wilderness of Philadelphia during the 1793 yellow fever epidemic, especially as it relates to the urban environmental implications of the disease itself.

Thus in this study I have avoided any such ranking and instead select texts that present themselves as important precursors or examples of the tradition of early American environmental fiction. I necessarily privilege those that take as their overt subjects the non-human world, animals, natural processes, and agriculture as coupled with environmental themes, but I do not forward these texts as the only, or in some cases even as the best, examples of environmental texts of their time. By ignoring Buell's earlier directive that for a text to count as environmental it must forward the idea of "*Human accountability to the environment*," we open for discussion a world of texts from the early republic that nonetheless merit an ecocritical take,[16] even if they often contain unapologetic accounts of greedy exploration, conquest, cruelty, wasted resources, and fanciful tall tales meant to promote the settlement of a place that was far from the paradise so many had claimed. With that criterion in mind, this book is meant to help address a less-studied but no less important facet of American literary history by focusing on the portrayal of the non-human elements of the natural world in works of early republic fiction.

---

[16] Buell, *The Environmental Imagination*, 7.

It is my hope that this book will complement a growing number of studies that deal with related matters. While interest in environmental criticism has grown significantly over the past couple of decades, most critical studies of American environmental literature begin deeper in the nineteenth century with Emerson's *Nature* (1836) and Thoreau's *Walden* (1854). Critics do at times make passing mention of early republic writers, but they tend to quickly dismiss those works in favor of later authors. Still, there are some notable critical studies that touch on the pre-Emersonian period, and I have found most of these of great value. For example, I have been particularly influenced by Rochelle Johnson's recent book, *Passions for Nature: Nineteenth-Century America's Aesthetics of Alienation* (2009), which stands as an illuminating examination of the paradoxes inherent in nineteenth-century environmental aesthetics. She makes especially important contributions to our understanding of Susan Fenimore Cooper's *Rural Hours* on the Transcendentalists, and her chapter on Thomas Cole provides a much-needed examination of the environmental lessons to be found in his more famous landscape paintings. Johnson's focus is on the period between 1836 and 1862, a brief but pivotal time in American environmental thought. She necessarily places most of her critical attention on Emerson, Susan Fenimore Cooper, and a handful of other influential figures, but many of her observations are applicable to earlier texts and American social trends. For example, she argues that in studying pre-1900 texts, "we uncover the richness of our cultural history, the various understandings of nature over time, and the implications of aesthetic renderings of the physical environment."[17]

A similar line of reasoning can be found in a number of other recent, and often diverse, ecocritical studies, each of which has much to offer regarding ways of valuing environmentally charged texts. For instance, Rinda West's *Out of the Shadow: Ecopsychology, Story, and Encounters with the Land* deals predominantly with contemporary texts, but at the core of her argument is the idea that

> human actions toward the natural world arise from assumptions about nature and the place of humans in nature that are deeply ingrained. These assumptions, which are usually not made explicit, are encoded in a variety of stories ... In particular, ideas about the relationship between humans and the rest of nature underlie most literary fictions.[18]

This same idea informs much of the reasoning in this book, that the fictions early Americans told themselves about the land did much to influence their treatment of that land.

Likewise, Robert E. Abrams's *Landscape and Ideology in American Renaissance Literature: Topographies of Skepticism* (2003) deftly investigates the intersections between an evolving popular conception of landscape and the

---

[17] Johnson, *Passions for Nature: Nineteenth-Century America's Aesthetics of Alienation*, 12.

[18] West, *Out of the Shadow*, 4.

controversies of national expansion, but, as its title indicates, the book focuses solely on mid-nineteenth-century literary culture. Thomas Hallock's *From the Fallen Tree: Frontier Narratives, Environmental Politics, and the Roots of a National Pastoral, 1749–1826* (2003) is a fine work of environmental literary and political history, especially as related to the oppression of Native Americans. Still, as Hallock focuses almost exclusively on the intersections between political and natural history, he does not directly address the literary nuances and evolution of the American environmental imagination or the fiction that sprang from it. Other books, such as Christoph Irmscher's *The Poetics of Natural History* (1999), offer insightful evaluations of early American natural history writing, but as helpful as these studies are, they do not tend to make a connection between those works of natural history and the nation's larger literary tradition, especially in terms of fiction. My study serves then to complement and build upon the work of these and many other critics by focusing on the development of environmental themes in the fiction of the day.

I argue that these themes appear in early works of U.S. fiction that offer a formative understanding of how humanity's influence alters the environmental status quo at the same time that the environment alters humanity. While present-day terms like *environmentalism* and *sustainability* were unknown to these writers, that did not stop the development of a proto-ecological consciousness that peeks from the pages of these early volumes. Given the lures of the Romantic age, it is unsurprising that this proto-ecological consciousness is often associated with pastoralism. It is, however, in its agrarian roots and pragmatic flavor markedly distinct from a merely romanticized Nature. It connotes instead a literal nexus between human subjectivity and what Emerson called the "NOT ME." The difference between the environmental vision of early republic fiction writers and the eventually more popular Transcendentalist spiritualism is the notion that the Emersonian "NOT ME" is itself a fiction. The writers in this study, each in their own ways, argue that humanity cannot so easily be extracted from the land upon which it depends for survival, even when that land seems so hostile to European colonization.

I open by examining how figures such as Crèvecoeur and Bartram created an imagined—and at times entirely fanciful—American environment that provided the essential foundation for the environmental fiction that followed. Crèvecoeur's *Letters from an American Farmer* (1782) and Bartram's *Travels* (1791) are major touchstones in the development of an American environmental consciousness. These writers were two of the first to pen serious studies of the American natural world and the human place therein, taking forward-thinking approaches by focusing on the complex communities of plants, animals, and people found in the American wilds. Specifically, Bartram creates a proto-ecological way of envisioning the environment, one in which he employs the tools of a fiction writer to create a dynamic portrait of the natural world and the human place within it. Looking beyond mere landscapes, Bartram promotes a complex understanding of natural communities that also incorporates the American Indians (namely, the

Muscogee [Creek], Choctaw, and Cherokee nations), who at that time were still the main human occupants of the region. In *Travels*, Bartram navigates a textual wilderness in which he composes his account through descriptions of biological processes meant not to capture static portraits of the landscape, but to illuminate living ecosystems ripe with vigorous biological communities. In *Letters*, Crèvecoeur also takes up the quill of a fiction writer, and like Bartram, finds that at times a little fiction helps make a better portrait of the American agrarian world.

The second and third chapters of this study examine how American fiction developed in tandem with the maturation of two forms of American popular literature, forms that are often overlooked in discussions of early environmental prose: Indian captivity narratives and juvenile literature. Specifically, I take a look at how that consummately American literary genre of the Indian captivity narrative incorporated an increasing environmentality while it was itself absorbed into the fiction of the late eighteenth century. In Chapter 3, I turn to how early American (and imported British) children's literature became progressively more environmental, especially in promoting the humane treatment of animals, the appreciation of the aesthetic beauty of the natural world, and, with the birth of the United States, the recognition that the nation's prosperity would rely upon judiciously exploiting its natural resources. I end Chapter 3 with a focus on the works of Samuel Griswold Goodrich, who, publishing in the 1820s and 1830s (often in his persona of "Peter Parley"), taught generations of children about a range of environmental topics touching upon the ethical, historical, and political aspects of the nation.

I devote the second half of this book to an examination of the environment as portrayed in three works by James Fenimore Cooper: *The Pioneers* (1823), *The Prairie* (1827), and *The Crater* (1847). I argue that each of these novels represent major accomplishments in the development of American environmental fiction. As necessary, I place these books in conversation with works by some of Cooper's contemporaries, in particular his father, William Cooper; James Kirke Paulding; and the famed landscape painter Thomas Cole. I begin with a chapter devoted to Cooper's *The Pioneers*, a book that in many ways serves as the core text of my study because it was the first fully realized and deliberate work of American environmental fiction. *The Pioneers* is remarkably prescient in its warnings about habitat destruction, about resource use, and about the need to learn to value the natural world for more than its economic potential. I am not alone in my esteem for the environmental depth of this book. Buell writes that "James Fenimore Cooper's *Pioneers* is a more faithful environmental text than the four ensuing Leatherstocking Tales because it never loses sight of the history of the community's development from wilderness to town."[19] Environmental historian Carolyn Merchant positions the book at the head of the American environmental literary tradition, pointing out that "*The Pioneers* … highlighted the ecological destruction of massive numbers of passenger pigeons, and sensitively portrayed

[19] Buell, *The Environmental Imagination*, 8.

the option of living by hunting only those animals needed for human survival."[20] This fact is made even more impressive when we remember that in 1823, when Cooper warned us through Natty Bumppo of the perils of overhunting, the now extinct passenger pigeon was still common to American skies. In my examination of this key novel, I argue that Cooper was not only a guiding force in developing an American brand of environmental fiction but that he was, himself, highly influenced by the genius of his father, William Cooper, for land speculation.

My fifth chapter focuses on how Cooper's third Leather-Stocking novel, *The Prairie* (1827), serves as a rebuttal to George-Louis Leclerc, Count de Buffon's theory of American degeneracy, which asserted that the American climate and lush vegetation created an environment that reduced the size and vitality of its mammals, including humans. Cooper, in this tale of violence and loss in an ethereal Great Plains landscape, refutes Buffon's theory by constructing his own myth of an American natural world populated, not with Buffon's diminutive weaklings, but with formidable animals and people who reside in a harsh, dry American prairie. Additionally, by satirizing European modes of biological inquiry via the comic character of Dr. Bat, *The Prairie* serves as Cooper's own contradictory volume of natural and social history.

I end with a study of the cultural utility of the natural disaster in the first half of the nineteenth century. From James Kirke Paulding's treatment of the massive New Madrid Earthquakes in his novel *Westward Ho!* (1832), to Thomas Cole's storm-riddled landscape paintings, to Fenimore Cooper's literally volcanic *The Crater* (1847), natural disasters play a significant role in the early imaginings of the natural world. This final chapter addresses how these novels variously complicated and reinscribed the conventional American understanding of natural disasters as acts of divine punishment (especially as it originated in the Puritan providence tale tradition). With these works, as well as others, such events begin to take on more complicated, if still punitive, significance.

Ultimately, I argue that early republic environmental fiction represents a valuable corpus of writings that say as much, if not more, about the invention of America as about the natural world itself. At the heart of this study is the idea that by better appreciating the development of our culture's ecological consciousness, by examining its false starts and first successes, we greatly enrich our present-day understanding of the natural world as a community of interdependence. Biologist Christopher Uhl, in describing our twenty-first century ecological consciousness, writes,

> The really big task ahead, it seems, is not to fix the environment but, rather, to fix our ways of seeing and, by extension, our ways of thinking ... *Fixing our ways of seeing* means coming to understand that Earth *is* our larger body .. You are, I am, each of us is a manifestation of Earth.[21]

---

20 Merchant, *American Environmental History*, 226.

21 Uhl, *Developing Ecological Consciousness*, 208.

We cannot approach this level of deep ecological understanding until we fully realize the basis of our old "ways of seeing"—our old ways of imagining—the environment. Complicit in this understanding is the recognition of the inevitably subjective character of our conception of the natural world and how that subjectivity is inextricably part of our language, our mythology, and our anxieties—including our desire to legitimize our national transgressions against humanity and nature alike.

The goal of this book is to contribute to the ongoing scholarly conversation about the origins, development, and direction of the environmental literary tradition of the United States. Environmental historian William Cronon writes, "An ecological history begins by assuming a dynamic and changing relationship between environment and culture, one as apt to produce contradictions as continuities."[22] Each text studied in this book serves as an artifact of this ever-developing relationship, of the invention of an American environmental consciousness, along with its attendant "contradictions" and "continuities." While this relationship has matured considerably over time, it has not followed a conveniently linear progression. After all, American fiction did not simply increase in ecological luminosity with each succeeding year. Instead, these texts appear as muted literary points, like stars emerging from a moonless dusk, gradually coming into view until finally, visible as a cluster of light, they illuminate the rise of American environmental fiction.

---

[22] Cronon, *Changes in the Land*, 13.

PART I
Verdant Beginnings

At the return of the morning, by the powerful influence of light, the pulse of nature becomes more active, and the universal vibration of life insensibly and irresistibly moves the wondrous machine. (161)

—William Bartram, *Travels* ... (1791)

Chapter 1
Imagining Natural Communities

> Benjamin overlooked NATURE. But the French Crèvecoeur spotted it, long before Thoreau and Emerson worked it up.
>
> —D. H. Lawrence, *Studies in Classic American Literature* (1923)[1]

I open my examination of early environmental fiction in the United States with two books that, especially given their influence on later writers, represent what I argue is the general beginning of the nation's imagined environment: J. Hector St. John de Crèvecoeur's *Letters from an American Farmer* (1782) and William Bartram's *Travels* (1791). That these books are presented by their authors as works of nonfiction demonstrates how the fictionalization of the nation's environment began with the transitioning from a predominantly non-fiction mode to a hybrid discourse that was symptomatic of the nonfiction writers' struggle to place into language an ultimately indescribable natural world. Finding no words to convey the magnificence of the environment before them, these and other authors began to incorporate elements from the realm of fiction, and in so doing blazed a trail for those works of outright environmental fiction that followed.

Each author was also concerned with much more than just providing an appealing description of the nation's natural beauty and agricultural promise. Bartram strives to impart a developed sense of ecological interconnectivity by promoting the idea that organisms constitute communities and that people are themselves intimately linked to, and reliant upon, the land. In the case of Crèvecoeur, he repeatedly employs elements of fiction, especially the fable or tall tale, to fulfill his goal of presenting us with a highly complex sociopolitical critique of the new republic, one that persuasively incorporates elements of the natural world as didactic tools in the teaching of morality and social responsibility.

Crèvecoeur's and Bartram's books have received a generous amount of literary analysis, but they remain fertile texts from which we may learn much about the beginnings of the new republic's environmental mind. With the aim of unearthing some new ways of looking at these seminal texts, I argue that each of these authors at times melded a relatively well-known nonfiction approach with a sensibility and techniques meant for fiction. In their hands the imagined environment became a more conscientiously employed fixture in nature writing. The result was a new breed of ecological narrative that served as the origin for the various works of environmental fiction that followed, works that were simultaneously about the natural and the national worlds—two major concerns in the early republican period.

---

1 Lawrence, *Studies in Classic American Literature*, 33.

Going in the order of their inception rather than their publication, I begin with a look at how William Bartram employs a fledgling environmental imagination to record and give life to the indescribable beauty and sublimity of the wilds of the North American Southeast as he found them in the late eighteenth century. My aim is not to argue that Bartram is absolutely the first to accomplish this task but that he did so in a way that is compelling and instructive of how later writers would deftly merge the actual natural world with an imagined one. Similarly, my discussion of Crèvecoeur's *Letters* looks at how he employs metaphor and allegorical anecdotes to better align his proto-conservationist ideal with the larger sociopolitical aim of the book. Each of these men write a form of fictionalized non-fiction, one that allowed them to build a more purposefully crafted narrative upon a foundation of actual experience. The product of this merger is a literary laboratory in which the authors take advantage of the powers of fiction while also firmly grounding their lessons in the situation in the real environment.

Bartram's Ecological Imagination

In his famous botanical garden bordering Pennsylvania's Schuylkill River, the prominent eighteenth-century botanist John Bartram taught his son William the relatively new Linnaean system for identifying and describing plants.[2] As Bruce Silver points out, William's botanical education, his artistic ability, and his father's reputation allowed him the opportunity to travel the wilds of the American Southeast collecting plants for Dr. John Fothergill. A wealthy London physician and gardener, Fothergill avidly collected botanical specimens and hoped to maintain his status as a collector of exotic plants by hiring Bartram to help supply his collection.[3] Between April 1773 and January 1778, Bartram walked, rode, and paddled through the southeastern wilderness collecting specimens, making drawings, and most importantly, keeping extensive field notes. These notes and a few of the drawings would—over a period of eighteen years—form the basis of one of the first serious studies of the American natural landscape, the dauntingly titled *Travels through North & South Carolina, Georgia, East & West Florida, the Cherokee Country, the Extensive Territories of the Muscogulges, or Creek Confederacy, and the Country of the Chactaws; containing An Account of the Soil and Natural Productions of those Regions, together with Observations on the Manners of the Indians*.

Like Bartram's landmark journey into the wilds of the Southeast, his written account of this expedition also enters new territory—a new imagined landscape the creation of which helped set the stage for the rise of American environmental fiction. Bartram navigates a textual wilderness in which he composes his account

---

2 John Bartram, botanist to King George III, counted among his friends Benjamin Franklin and Benjamin Rush.

3 Silver, "William Bartram's and Other Eighteenth-Century Accounts of Nature," 597.

through descriptions of biological processes meant, not to capture static portraits of the natural world, but to illuminate a living landscape—a landscape ripe with vigorous ecological communities. These natural communities also incorporate humans in the form of American Indians who, Bartram's narrative suggests, represent the proper way to thrive in the wilds of southeastern America. *Travels* represents a shift in biological thought from single-species studies to examinations of multiple species and their interactions within specific ecosystems, and this shift not only required Bartram's keen biological insight, it also required his imagination. To really capture the dynamism and complexity of the natural world, he repeatedly creates an ecological narrative, one in which the components of an environment are set into action with each other. Unlike any naturalist before him, Bartram used the elements of good storytelling to tell the tale of the land.

Though 75 years would elapse between the 1791 publication of *Travels* and Ernst Haeckel's coining of the word *oecologie*,[4] Bartram and other writers such as Gilbert White were already looking beyond static natural description to write about nature as a complex system of interaction among living organisms. This awareness led to a holistic view of the natural world. While White commented on the domesticated countryside he found in England, Bartram wrote of a much different nature. Home to alligators, rattlesnakes, and mysterious American Indians, the southeastern American wilderness Bartram encountered had not yet been tamed by Europeans. Bartram found an extremely diverse natural environment that defied conventional modes of description. Neither pastoral poetry nor scientific jargon could adequately describe the wilds of the American Southeast. To compensate for the difficulty of describing a new world in old terms, Bartram's narrative hybridizes conventional descriptive strategies. He creates a confluence of pastoral imagery and biological description that develops into an imaginative mechanism for relating cyclical biological communities.

> How gently flow they peaceful floods, O Alatamaha! How sublimely rise to view, on thy elevated shores, yon magnolian groves, from whose tops the surrounding expanse is perfumed, by clouds of incense, blended with the exhaling balm of the liquidambar, and odours continually arising from circumambient aromatic groves of illicium, myrica, laurus, and bignonia.[5]

As this passage demonstrates, Bartram's *Travels* melds the scientific and the pastoral.[6] Beginning with a poetic apostrophe to the Alatamaha River, his description soon transforms into a list of taxonomic names for the trees along the bank. The trees in this description occupy the environment defined by the river;

---

[4] The International Botanical Congress changed the spelling of *oecologie* to *ecology* in 1893. See Worster, *Nature's Economy*, for a detailed history of ecological thought.

[5] Bartram, *Travels*, 63. Subsequent references appear parenthetically in the text.

[6] Glen A. Love in *Practical Ecocriticism* examines how some scientists, such as E.O. Wilson, with his idea of biophilia, have investigated the evolutionary science behind humanity's pastoral impulse; see especially 72–88.

therefore, they represent a description, not only of some trees, but of an ecological community. In this merger, Bartram recognizes "odours continually arising from circumambient aromatic groves" (63). The suspected permanence of these natural odors indicates prolonged biological stability. Indeed, this description hints at his recognition of homeostasis, the now controversial idea that healthy ecosystems tend to maintain a stable relationship with their components and surroundings.

During the last few decades the critical response to *Travels* has moved from analyses of Bartram's influence on the works of writers such as Samuel Coleridge to studies concerning a number of its environmental contributions to the American literary tradition.[7] The varied criticism reflects the diverse nature of *Travels*. It is not an overstatement to say that *Travels* is a botanical, zoological, anthropological, ethological, agricultural, theological, sociological, and literary work of great importance. Critical responses also comment on the book's political agenda, on Bartram's aesthetic philosophy, and especially on his descriptions of the American Indians he frequently encountered.[8] His descriptions of American Indians are surprising because he did not proliferate the racist stereotypes so prevalent in his day, and for decades to come.

Despite this healthy body of critical attention, relatively few critics center on *Travels's* actual ecological and literary importance. One exception is Charles Adams, who persuasively argues that *Travels* embodies the first work in a long line of American literary nature writing, a tradition that includes such classics as Henry David Thoreau's *Walden*, Aldo Leopold's *A Sand County Almanac*, and Rachel Carson's *The Sea Around Us*. Adams contends that

> Like all self-consciously literary nature writing, Bartram's *Travels* is more about the writer than about nature ... Bartram and his fellow nature writers insist that humanity's relationship with nature is fundamentally a matter of articulated consciousness—fundamentally, that is, a matter of language. The story the nature writer has to tell concerns the process of turning the natural world into words, with language the focus of the writer's relationship with her or his subject.[9]

Taken from this perspective, Bartram's fundamental goal becomes, not solely to record natural history, but also to evoke in the reader, through the representation of ecological communities, his experience of the southern wilderness. Not all critics agree that Bartram is successful in this endeavor. Pamela Regis argues

7 For more about Bartram's influence upon Coleridge, see Lowes, *The Road to Xanadu*.

8 For a discussion of Bartram's politics, see Anderson, "Bartram's *Travels* and the Politics of Nature," and Looby, "The Constitution of Nature." L. Hugh Moore, in "The Aesthetic Theory of William Bartram," provides an excellent analysis of Bartram's aesthetic ideals; for a detailed discussion of Bartram's representations of American Indians, see Joshua Bellin, "Wicked Instruments" and Waselkov and Braund, *William Bartram on the Southeastern Indians*.

9 Adams, "Reading Ecologically," 67.

that Bartram's narrative fails because "In his descriptions America is a vast, still garden, planted and denizened with species that have names but no one to use them. It is known, but not truly inhabited."[10] While the narrative of *Travels*—like much natural history writing—does lack a dynamic and conventionally entertaining narrative, to say that it represents a world "not truly inhabited" fails to account for Bartram's representation of natural communities. It also underserves the value of Bartram's imagined environment itself. After all, if we are to concede Adams's point that *Travels* is much more an account of its author's perception of the natural world than an actual record of the natural world itself, then divergences from reality are unavoidable.

In *Travels*, Bartram tells the story of the entire wilderness, not of only a few species. To accomplish this ecological agenda, he must widen his environmental imagination by populating the narrative with collections of individuals contained by both geological and biological boundaries. These boundaries constitute a complex network of living ecological communities. In his book *Community Ecology*, ecologist Peter J. Morin writes that community ecology is "the study of patterns and processes involving at least two species at a particular location. This broad definition embraces topics such as predator-prey interactions and interspecific competition."[11] Modern community ecologists study concepts such as trophic levels and food chains, terms that refer to the feeding habits of organisms within a particular ecosystem. While many ecologists credit the work of early twentieth-century biologists such as F.E. Clements and V.E. Shelford with creating the field of community ecology, Bartram devotes much of *Travels* to these same issues. As his narrative repeatedly indicates, Bartram was quite aware of ecological concepts even if he lacked the terminology currently used to discuss them. His profound appreciation for the unique character of the southeastern ecosystem may even mark him as one of the first to write about the natural world in terms of biotic communities, mutualism, and succession. As he writes in the introduction to *Travels*, "If we bestow but very little attention to the economy of the animal creation, we shall find manifest examples of premeditation, perseverance, resolution, and consummate artifice, in order to effect their purposes" (21). He engaged in a diversity of figures of speech and imaginative conceits drawn from botanical, theological, and economic thought, all of which he used to mirror the complexity interwoven into the landscape of the southeastern wilderness. Adams discusses these techniques, commenting that *Travels* "is characterized by a complex, and ironic play of rhetorics" that convey the experience of the environment.[12]

Bartram's narrative techniques ensure that *Travels* succeeds where dried botanical specimens and cold descriptions of individual organisms fail. On numerous occasions throughout the text, the reader experiences the southeastern wilderness of the 1770s through thoughtfully designed representations of

---

[10] Regis, *Describing Early America*, 78.

[11] Morin, *Community Ecology*, 6.

[12] Adams, "Reading Ecologically," 70.

ecological communities. One of the most cited passages of *Travels* illustrates this concept, particularly as it relates to current ecological study of predation and energy dynamics. In observing a swarm of mayflies, which he calls by their taxonomic name, Ephemera,[13] Bartram sets the scene:

> Our place of observation is happily situated under the protecting shade of majestic Live Oaks, glorious Magnolias, and the fragrant Orange, open to the view of the great river and still waters of the lagoon just before us.
>
> At the cool eve's approach, the sweet enchanting melody of the feathered songsters gradually ceases, and they betake themselves to the leafy coverts for security and repose. (87)

Here he provides a quick overview of the ecological setting; he adequately describes the species of the trees and their proximity to a lagoon. He also records the time of day and the presence of various birds. This description boasts both a poetic flavor and a scientific attention to detail. Taken together these two descriptive approaches construct a rough outline of the biotic community. He next focuses on the true subject of his observation:

> Solemnly and slowly move onward, to the river's shore, the rustling clouds of the Ephemera. How awful the procession! Innumerable millions of winged beings, voluntary verging on to destruction, to the brink of the grave, where they behold bands of their enemies with wide open jaws, ready to receive them. (87)

Bartram's narrative introduces ecological interaction within the community of river, trees, and, now the mayflies. With a dose of anthropomorphism making "enemies" of those animals that feed upon the swarms of mayflies, he signals that the specific type of biological interaction is predation—a phenomenon he returns to many times throughout *Travels*. Engaging in further anthropomorphism he next introduces another key part of the ecological picture:

> But as if insensible of their danger, gay and tranquil each meets his beloved mate in the still air, inimitably bedecked in their new nuptial robes. What eye can trace them, in their vertical wanton amorous chaces, bounding and fluttering on the odoriferous air! With what peace, love, and joy, do they end the last moments of their existence? (87)

Reproduction enters the scene and together with the descriptions of predation completes two dramatic stages of the life cycle of the mayfly, doing much to help illustrate the complex community of the area. Notably, in creating this scene Bartram must resort to imaginative, if decidedly unscientific, instances of anthropomorphism—a tool he uses throughout the text. In his study of anthropomorphism in American literature, Bryan L. Moore notes Bartram's

---

[13] The Ephemera are hereafter referred to as mayflies.

affection for this literary technique, writing that sometimes his "pathetic fallacies seem strained to a modern reader … Yet Bartram is no mere pastoralist."[14] After his initial description, Bartram employs yet another strategy to help explain the gravity of the scene:

> The importance of the existence of these beautiful and delicately formed little creatures, whose frame and organization are equally wonderful, more delicate, and perhaps as complicated as those of the most perfect human being, is well worth a few moments contemplation ... what a lesson doth it not afford us of the vanity of our own pursuits. (88)

He underscores the significance of the organism in human terms and frames the lifecycle of the mayfly as a "lesson" from which the reader might learn about the human condition within the natural world. This approach brings humans into juxtaposition with the components of the described community. Bartram places the reader directly within the influence of the natural scene he describes, asking the reader to interact with the mayflies in much the way the birds do, by devouring their presence for a form of betterment. This particular connection between humans and mayflies suggests his holistic view of nature. This understanding of natural interconnectedness leads him to avoid constructing boundaries between humanity and the natural world.

His belief in an interconnected world might best be demonstrated in his portrayal of the American Indians as people from whom European Americans can learn much about living in nature. *Travels* showed its European and European American readers the rarest of American Indians—ones who did not embody the racist stereotypes prevalent at the time. In fact, Bartram spends a significant portion of *Travels* detailing his encounters with the Indians, and for the most part, he presents them as people who in their actions and culture appear similar to European Americans. Sometimes he does fall prey to the noble savage stereotype, but he never portrays Indian in the devils-of-the-forest guise. He shows respect in his attempts to accurately record the various aspects of Indian culture. Hoping to help preserve this misunderstood human segment of the natural community, Bartram devotes the remaining quarter of *Travels* to bringing the particulars of Indian culture to European and European American readers. As Gregory A. Waselkov and Kathryn E. Holland Braund contend, "'Some Hints & Observations' is primarily an eloquent plea for equal treatment of Native Americans as brothers under God. Most importantly, this work sets forth—in Bartram's view—the incontrovertible claim of the Natives to the land they occupied."[15] Bartram's interest in the Native Americans centers on their ability to successfully occupy their wilderness home.

Yet his portrayal of the Native Americans is complicated by political and cultural biases that ultimately indicate the fictions inherent in the text. In discussing Bartram's portrayal of the political and economic borderlands between

14 Bryan L. Moore, *Ecology and Literature*, 106–7.

15 Waselkov and Braund, *William Bartram on the Southeastern Indians*, 188.

a wilderness controlled by Native Americans and destined for European American colonialism, Thomas Hallock reveals how *Travels* stands as a highly constructed text. This construction approaches that found in a work of fiction with an edited plot and characters that have been modified for dramatic or persuasive effect. For example, early in the book Bartram has a tense but friendly encounter with an intimidating Seminole man whom he later discovers was "a noted murderer" ostensibly out to "kill the first white man he met" (15–16). Of this scene Hallock writes, "The episode plays a fairly clear role in the plot. Bartram's editors note that the encounter occurred much later in real time; by folding a brush with death into the opening pages, the author establishes the theme of native-white relations against the botanic self."[16] By modifying the events and highlighting characterizations of Native Americans Bartram reveals the level to which this influential book was edited to support a rationalization of the colonialism of the area.

Indeed, upon an initial reading, *Travels* appears to hinge upon an ambiguous land use philosophy of natural appreciation and prospective utilization of the environment. Bartram simultaneously praises the aesthetic beauty of the wilderness while suggesting its possible development:

> This is, perhaps, one of the most eligible situations for a city in the world; a level plain between the conflux of two majestic rivers, which are exactly of equal magnitude in appearance, each navigable for vessels and perriauguas at least five hundred miles above it, and spreading their numerous branches over the most fertile and delightful regions. (360)

For him, this mixed agenda does not represent a mutually exclusive condition because he views humans as part of the natural world, as members of an ecological community. His narrative often portrays human use in the context of the natural world. Though he may traverse a wilderness, it is populated and reveals a history of human use presented in his descriptions of the many ruins he encounters, both Indian and European. Bartram finds a particularly good example during a wedding ceremony at a Creek town where

> a booth or pavilion was formed of green boughs, having two Laurel trees planted in front (Magnolia grandiflora). This was the secret nuptial chamber. Dancing, music and feasting continued the forepart of the night, and towards morning the happy couple privately withdrew, and continued alone all the next day, no one presuming to approach the sacred, mysterious thalame. (359–60)

Following this account of a merger not only between two people but also between nature and culture, marriage ceremony and magnolia, Bartram pans out to tell another story, one of a failed marriage:

> At the confluence of the two fine rivers, the Tallapoose and Coosau, which here resign their names to the great Alabama, where are to be seen traces of the

---

16 Hallock, *From the Fallen* Tree, 159.

> ancient French fortress, Thoulouse; here are yet lying, half buried in the earth, a few pieces of ordnance, four and six pounders. I observed, in a very thriving condition, two or three very large apple trees, planted here by the French. (360)

He overtly advocates the use of the area, but his presentation takes the form of a thinly veiled lesson: unlike the disintegrating French fortress, the Indian town embodies a natural extension of the surrounding wilderness. Here the Creeks build their nuptial chamber with "green boughs" around two existing magnolias, the French "planted" their legacy, leaving "half-buried in the earth" artillery pieces and "thriving" apple trees. The ruined fortress and the cannons half buried in the ground represent an erroneous approach to living in the wilderness, while the two apple trees—in their natural glory—embody the only lasting portion of the French legacy. The point becomes clear even without the possibly coincidental allusions to Eden: European Americans must not build fortresses between themselves and the broader natural community; they, like the Creeks, must incorporate—in significant and sustaining ways—the natural world into their own cultural practices. Bartram's moral is that only those humans who accept their inextricable connection to the natural and dynamic environmental community stand a chance of surviving.

Trained as a botanist, Bartram's understanding of dynamic communities exists in those numerous parts of his narrative devoted to taxonomic descriptions of the plants and animals. In one instance, after a storm forces him to land his small boat on an island in a large lake, he goes about, not only describing the features of the location, but also constructing a complex community. Notably it is when he levies his imaginative powers in tandem with his taxonomic expertise that he creates instructive fictions of the natural world, tales of community interaction that while almost always reliant upon invocations of the pathetic fallacy and flights of fancy, nevertheless result in a much more ecologically instructive vision of the natural world than possible through science alone. For example, he speculates the island "from obvious vestiges, to have been once the chosen residence of an Indian prince, there being to this day evident remains of a large town of the Aborigines" (103). This passage, like many in *Travels*, begins with Bartram's description of the island's human history shown through the vestiges of a past civilization. After further describing the ruins of this abandoned Indian town, he moves into the botanical realm, "through a magnificent grove of magnolias, live oaks, palms and orange trees, terminating at the verge of a large green level savanna" (103). Following this dendrological examination, Bartram comments on the "great number of deer, turkeys, bears, wolves, wild cats, squirrels, raccoons, and opossums" (103). This general list of animals joins to the ecological community, adding characters to the narrative.

Immediately following his introduction of the animals, Bartram briefly sets a small portion of the narrative community in motion by writing that "The bears are invited here to partake of the fruit of the orange tree, which they are immoderately fond of; and both they and turkeys are made extremely fat and delicious, from their feeding on the sweet acorns of the live oak" (103). For a moment the bear and turkey interact with the orange trees and live oaks of the island. By alluding to human use in describing the "fat and delicious" aspect of the bears and turkeys, the community extends to incorporate the hunter who visits the island, like the bear, to obtain food. The narrative includes man in the natural community, and while this addition may appear to violate the borders of the natural system, Bartram, as the constant references to human influence suggest, does not necessarily separate the two. This segment of the narrative ends with his detailed description of several plant species and the songs of shore birds. The various types of description cycle between humans, plants, animals, and features of the landscape, culminating in an ecological community. He constructs a scene populated by diverse organisms that function together within a specific location. The island in this example functions as a small sample of what Bartram accomplishes throughout *Travels* with the use of a stratified series of descriptions, actions, analogies, allusions, and numerous other literary techniques.

As Regis argues, Bartram's use of description—particularly his exhaustive lists of taxonomic names—"provides a local habitation or context ... The list represents to the reader an American scene."[17] But Regis's claim that the narrative "fails to provide a middle ground between the impersonal facts of natural history and the psychologically immediate sensations of Burkean aesthetics" assumes that he perceived a middle ground at all.[18] As Charles H. Adams writes, "Linnaean nomenclature and Bartram aesthetic tropes are only two of the many varieties of rhetoric that Bartram weaves throughout the *Travels*. They range from the descriptive, through the metaphoric and moral, to the religious."[19] His accumulation of these various rhetorical techniques culminates in a rich narrative of ecological community. Each literary element builds upon the other to form a picture of what no single component conveys on its own, a portrait of the southeastern American wilderness.

Bartram was also an accomplished illustrator who occasionally relied upon literal portraits to communicate his experience of the natural world. Often these drawings reflect simply the standard, detailed, single-specimen studies so commonly found, even today, in botanical and zoological works. Perhaps the most famous example is his drawing of *Franklinia alatamaha* (Figure 1.1),[20] a rare tree that John and William Bartram discovered in 1765 while in Georgia.

Naming the tree after John Bartram's friend Benjamin Franklin, William drew it in the great detail customary as part of competently recording the description of a plant specimen. The drawing of *Franklinia alatamaha* shows the tree's leaf structure, fruit, seeds, and especially the all-important floral characteristics needed to describe the plant in Linnaean terms. Ironically, the illustration, like most of

---

17 Regis, *Describing Early America*, 60.

18 Regis, *Describing Early America*, 42.

19 Charles H. Adams "Reading Ecologically," 70.

20 For other examples of Bartram's many drawings see Bartram, *Botanical and Zoological Drawings, 1756–1788*.

Fig. 1.1 William Bartram (1739–1823), *Franklinia alatamaha*, 1788. (015901). © Natural History Museum, London.

this type, never actually shows a representation of the entire tree. The drawing presents the tree completely outside of its natural context. Like a cutting on a table, the picture shows the viewer only the portions necessary to identify and study the individual plant.

Other drawings show Bartram's understanding that a dynamic representation of a living community boasts superior qualities to a simple, sterile, non-contextualized illustration. One of the most dramatic of these drawings of an imagined ecological community in action is *The Sarasena*—a title referring to the genus name of the pitcher plant (Figure 1.2).[21] Thomas P. Slaughter writes that *The Sarasena* "is William's most striking image of all nature as somebody's meal. The picture is most effective if you start looking at the lower left-hand scene to make our place clear. Mankind, the great hunter of all, becomes in our turn the prey of insects and worms. The circle is full."[22] This "circle" represents the cyclical process of ecological communities, a chain of devourers. Yet this drawing is much more than simply a picture of organisms eating each other. Each organism presented occupies the same location and therefore they represent interaction. This dynamic drawing encompasses much more than the simple picture of *Franklinia alatamaha*. Rather than a cold portrait that lacks context, Bartram delivers an action pose, albeit idealized, of a natural community. Illustrations such as *The Sarasena* tell an ecological story of predation, reproduction, and survival based on, and defined by, biological interactions.

Bartram's ecological narrative depends heavily upon the role of the predator in the natural community. His descriptions view the violence of predation as key to the continuation of the ecological cycles that tend to preserve communities. One predator of special interest, the insectivorous plant, fascinated Bartram, who often included pitcher plants and Venus flytraps in his illustrations. In *The Sarasena*, the pitcher plant stands directly opposite a scarlet snake busily devouring a frog; this juxtaposition presents two views of predation. The pitcher plant not only derives its meals from trapping unfortunate insects, it also, like most plants, pulls a large amount of nutrition from the soil. While Slaughter mentions insects and worms as the devourers of man, plants also fulfill this role. For Bartram, the pitcher plant represents a consummate devourer and as such embodies a particularly fascinating niche in the biotic community. Slaughter points out that Bartram's nature resides in a state of constant conflict and displays "a clear-eyed knowledge of the harshness of nature, an acceptance of the violence and pain that were as natural as the serenity of a summer's blue sky ... all consumed all in William's wilderness, in a relentless warfare among animals and plant."[23] As this description of the Venus

---

[21] For convenience, I use Slaughter's title (in *The Natures of John and William Bartram*) for this particular drawing. The image's full title is *Sarracenia flava, yellow pitcher plant & Sarracenia purpurea, pitcher plant.*

[22] Slaughter, *The Natures of John and William Bartram*, 197.

[23] Slaughter, *The Natures of John and William Bartram*, 197.

Fig. 1.2. William Bartram (1739–1823), *Sarracenia flava, yellow pitcher plant & Sarracenia purpurea, pitcher plant—The Sarasena,*" circa 1773–1777. (015930). © Natural History Museum, London.

flytrap suggests, for Bartram, insectivorous plants ideally embody this concept of constant opportunistic conflict:

> admirable are the properties of the extraordinary Dionea muscipula! ... Astonishing production! see the incarnate lobes expanding, how gay and sportive they appear! ready on the spring to intrap incautious deluded insects! what artifice! there behold one of the leaves just closed upon a struggling fly; another has gotten a worm; its hold is sure, its prey can never escape—carnivorous vegetable! Can we after viewing this object, hesitate a moment to confess, that vegetable beings are endued with some sensible faculties or attributes, similar to those that dignify animal nature; they are organical, living, and self-moving bodies, for we see here, in this plant, motion and volition. (17)

Bartram uses the sensationalism of insectivorous plants to prove a point about all plants in general, that they—like animals, like people—play a major and complex role in the nutritional community. As this passage illustrates, much of the power of Bartram's description comes from his creation of a series of miniature stories that put the plants and insects, his characters, into motion on the page. Appearing to Bartram as a "gay and sportive" plant, the Venus flytrap springs its trap upon its insect prey, and through this literary strategy he makes a point to display the "motion and volition" of these plants as something akin to an uncanny scene in the never-ending series of tiny dramas that take place upon the environmental stage.

Readers who see Bartram's wilderness as simply a place of unending struggle and pain miss the overall effect of the conflict. Along with the violent images of a snake eating a frog and of a lizard preparing to capture a cricket another, more complex, image emerges. This impression lends significance to the natural warfare and partially explains the purpose of such suffering. In *The Sarasena*, one portion of the pitcher plant leans upon its side, its "lid" open, partially revealing a fruit of some type along with a small flower. This image—especially when viewed in conjunction with the nautilus shape of the snail shells—suggests a cornucopia, a source of infinite prosperity. In his introduction to *Travels*, Bartram relates this exact image in his description of the pitcher plant, "the hollow leaves are not unlike the cornucopia or Almalthea's horn; what a quantity of water a leaf is capable of containing, about a pint! taste of it—how cool and animating—limpid as the morning dew" (16). Thus, *Travels* and *The Sarasena* are powerful representations of imagined ecological communities that, while often violent, also enjoy moments of "limpid" tranquility.

Crèvecoeur's Allegorical Nature Writing

Written by a Frenchman, about America, and first published in London in 1782 (the year Britain effectively conceded defeat to the American colonists), Crèvecoeur's *Letters from an American Farmer* occupies a unique space in the early republic's literary canon. Like Bartram's *Travels*, the book is variously fiction and nonfiction, social commentary and local color, proclaiming America's fledgling nationalistic identity to the people of Europe. As such, it is an account of the land and people associated with the exotic new nation, but it is also in many ways an outright work of fiction, including its fictionalized narrator, the provincial farmer James, who despite his initial reluctance to write the letters agrees and even relishes his task of interpreting America and its ways to a presumably European readership. While Bartram's field notes sat for years waiting for him to put them into book form, Crèvecoeur penned one of the most influential and difficult to categorize books in U.S. literary history.

Prescient in its attention to the young United States, *Letters* offers insights into the complexities inherent in the character of both Americans and America itself. The book's fictitious narrator, a literary persona in the guise of a New York farmer named James, famously asks, "What, then, is the American, this new man?"[24] And in answer, Crèvecoeur wrote a book as much about the natural world as about those who worked to carve a nation from it, as much about the land and sea as about those who cultivated it, fished it, and called it home. Just as Crèvecoeur's book straddles and blurs nationalistic lines (Is the book a product of America, or is it merely about America?), so too does it frustrate attempts to categorize it in terms of genre. While this epistolary work (as well as Crèvecoeur's other pieces that eventually made up his *Sketches of Eighteenth-Century America*) seems to belong in the realm of nonfiction, it quickly betrays profound fictionalizations. Some of these tall-tales far exceed the license taken in Crèvecoeur's invention of a narrator and his consummately American farm family. The liberties the book takes with objective truth as well as the sometimes-sensational character of its narrative have encouraged some critics to label it nothing less than a novel. Dennis D. Moore comments on this trend by writing, "Many critics tend to look to Crèvecoeur for evidence of the first American prose fiction, if not for the first novel in America," and Moore offers one example in Albert E. Stone, who, in his introduction to the Penguin edition of *Letters*, asserts that Crèvecoeur "is by predilection an embryonic novelist."[25] Grantland S. Rice contends that "*Letters* is best read as an early but sophisticated novel."[26]

---

24 Crèvecoeur, *Letters from an American Farmer* and *Sketches of Eighteenth-Century America*, 69. Subsequent references appear parenthetically in the text.

25 Dennis D. Moore, Introduction in *More Letters from the American Farmer*, xxxvii; Stone, Introduction in *Letters from an American Farmer*, 22.

26 Rice, "Crèvecoeur and the Politics of Authorship in Republican America," 93. For other critics that support, with variations, the status of *Letters* as a novel, see Holbo, "Imagination, Commerce, and the Politics of Associationism in Crèvecoeur's *Letters from an American Farmer*" and Bauer's *The Cultural Geography of Colonial American Literatures*, in which he argues that Crèvecoeur in writing the persona of James creates a "narrative mask of self-effacement assumed by the fictional American type," and that "The value of James' testimony ... derives in part from its 'authenticity' as oral speech ... The Fiction of James' early letters as transcribed oral language is intensified by his short sentences, broken up by semi-colons, and his paratactic style," 211, 213.

Simply labeling *Letters* as a novel, however, introduces the danger of oversimplifying what is a remarkably sophisticated work of sociopolitical and agricultural/environmental commentary. At the core of the problem is determining how to separate the author from his narrative persona of Farmer James. As Ed White cautions, "The reading of the *Letters* as 'novelistic' was an important moment for Crèvecoeur scholarship. In emphasizing literary qualities, above all character and narrative trajectory, a much needed irony rescued the text from an earlier tendency to conflate James and his creator, while reading the narrative as so much sincere and naïve data." However, as White warns, "Once the character of the 'farmer' is reduced to a literary device denoting the epistemological privileges of the New World, Crèvecoeur is decisively shifted away from the agrarian sphere to that of the cosmopolitan intellectual. In some sense, then, Crèvecoeur remains conflated with James, who is simply a vernacular *philosophe*."[27]

One challenge, then, remains the need to understand *Letters* as a novelistic work, but one that bears the hallmarks and objectives of the very non-fiction it parodies and absorbs, most especially that of natural history. Perhaps nowhere in the book is Crèvecoeur's odd mixture of fact and fiction more pronounced than in those passages where he writes about animals, especially wild animals. At times his—or more accurately James's—portrayals of animals veer so deeply into the territory of folklore and tall tales that it is difficult to understand how his contemporary readers could have taken him seriously. For example, few Americans would have believed James's claim that a man bitten by a copperhead would then take on the behavior of a snake before dying. But Crèvecoeur probably did not much care about the decidedly small American readership. He was out to attract European readers, as they were essentially the only viable audience for such books. The realities of the printing market required that the book be published in Europe anyway. America simply did not have the printing infrastructure. Still, as unfamiliar as European audiences were about the ways of American animals, it is a stretch to think that he intended to portray such creatures merely for the purpose of sensationalism. The allegorical character of so many of his animal descriptions suggests that Crèvecoeur had little intention of duping the European crowd with weird animal tales. He instead used these various fanciful representations allegorically, as a means to employ the imagery of American wildlife to illuminate the prevailing sociopolitical and philosophical mindset he found in the formative nation.

Crèvecoeur telegraphs this intention early in *Letters* when he has James relate one of the most cited animal stories of the book: that of the kingbird and the bees. An avid beekeeper, James leads into his story by citing a number of instances of metamorphosis, and of the power of humanity to control these transformations. He notes how allowing his son to ride upon the plow as he works his fields pleases the little boy, and he "looks more blooming since I have adopted that practice" (55). Furthering this line of thinking, Crèvecoeur's narrator then follows with several

27 White, "Crèvecoeur in Wyoming," 401.

more examples, each of which tie in with his recognition of the transformative qualities of the natural world and of the human ability to appreciate, ignore, subvert, or stop such natural occurrences. He writes of returning from the fields at the end of the day to see "myriads of insects ... dancing in the beams of the setting sun" and how they are "carefully improving this short evening space" (55). James then shifts from the aesthetic appreciation of these insects—a transformation of the insects into beautiful dancers—to more pragmatic meditations on the wonder of interrupting natural moments of metamorphosis. He writes, "I never see an egg brought on my table but I feel penetrated with the wonderful change it would have undergone but for my gluttony," and he admits, "I never see my trees drop their leaves and their fruit in the autumn, and bud again in the spring, without wonder" (55). With this assemblage of natural and human interrupted transformations of nature's processes, James constructs a natural world that serves as a constantly moving, constantly recycling mechanism of generation, degeneration, and decay. Mankind's gift is the ability to interfere with this mechanism as he sees fit, and though he is not godlike enough to stop the seasons, he can figuratively transform his son into a plant and insects into dancers; and as his example of the eggs reminds us, human power resides as much in the ability to stop these processes as to foster them. With these examples placed at the opening of James's tale, Crèvecoeur establishes the idea of mankind's power over natural transformations and even (as with the dancing insects and the termination of future chickens he sees in his eggs) of imposing transformations of his own, whether figurative or literal.

By the time he gets to his story of the bees, we are fully steeped in this reasoning, looking for natural changes and the human hand in relation to them. Crèvecoeur then begins a new transformation of his own, one directly related to James's bees. James continues his discussion of the wonders on his farm, and argues,

> the tenants of my farm astonish me; some of them seem to surpass even men in memory and sagacity ... What, then, is this instinct which we so debase, and of which we are taught to entertain so diminutive an idea? My bees, above any other tenants of my farm, attract my attention and respect. (55)

He notes that the voraciousness of another resident of this farm, the often beneficial king birds, had become especially damaging to his prized bees. He decided to kill these birds to discourage their predation upon the bees, but he writes that before he can fire a shot he witnesses an amazing sight: "a bunch of bees as big as my fist issued from one of the hives, rushed on one of these birds" (56). After their initial attack on their oppressor, the bees "quitted their military array and disbanded," a tactical error that allowed the king bird to recover from the attack to such a degree that he "snapped as many as he wanted; nay he had even the impudence to alight on the very twig from which the bees had driven him" (56). Following this mock epic battle, James kills the king bird and cuts open its craw. He writes that he found inside the bird exactly "171 bees," which he placed in the sun, whereupon "54 returned to life, licked themselves clean, and joyfully went back to the hive, where they probably informed their companions of such an adventure and escape as I believe had never happened before to American bees" (56). When discussing the prevalence of such anthropomorphism in early American literature, Bryan Moore argues, "Like rhetoric itself, personification is not the sole property of any given point of view or ideology; it is not an inherently preservationist or ecocentric trope."[28] But for Crèvecoeur it serves as a particularly effective narrative tool, especially in those moments when he employs it in the service of an allegorical lesson.

This much-studied anecdote seems an altogether allegorical narrative except for the fact of its ambiguity. Bearing all the hallmarks of an allegory but lacking a clear assignation of moral value to the players, we are left with more questions than answers. Why, for example, is James so specific about the number of bees eaten by the king-bird? This number not only seems extraordinarily high but also the result of a painstaking and unrealistic process of counting the consumed bees. Is James's time so free that he can spend it this way? Why, also, did 54 return to life? And why are these bees (which are clearly European honey bees) referred to as "American bees"? While no critic has satisfactorily answered these questions, it seems too fanciful a tale to be taken at face value. Crèvecoeur engages, though perhaps ineffectively, in telling an animal fable, one that seems to beg for an allegorical reading, but that in its telling rejects any clear analogies. The rationale behind the number of the bees remains a mystery, but Crèvecoeur seems to employ these elements of the natural world to comment upon the larger, actual battles then raging across the land. And we might register the message of this allegory as one indicating the importance of maintaining unity during crisis, as well as the idea that even when these "American bees" seem to have suffered defeat, they will eventually prevail and return home to tell their tale.

In analyzing this famous passage, Albert E. Stone writes that

> Crèvecoeur's characteristic utterance is through metaphor and fable; it is parable moving in the direction of myth. One notices, first of all, the anecdote's deliberately factual tone, lulling the unwary reader into accepting everything at face value. ... Only by the last sentence do we begin to suspect that the author has been pulling our leg.[29]

As Stone points out, while this fable is one in a tradition of American tall tales, its seriousness of subject precludes the kind of humor found a century later in Mark Twain's "stretchers." In fact, as Jeff Osborne has noted, the allegory seems to form part of Crèvecoeur's idea of nature's inherent cruelty and the need for individuals to bond together into social contracts that help promote their survival. Osborne writes,

> The king-bird and the bee anecdote ... might be read as a social-contract allegory: the bird as anthropomorphized enemy provokes the bees into solidarity ... The

28 Bryan L. Moore, *Ecology and Literature*, 108.

29 Stone, "Introduction," *Letters from an American Farmer*, 20.

> illustration conforms to the philosophical narrative of social contract: strife and the threat of violence induce the bees to enter into an unspoken contract, ceding their personal power to a collective will that offers protection from such threats.[30]

In the end, the tale of the king-bird and the bees makes for a vague and perplexing pseudo-allegory, one particularly enigmatic when considered within the content of Crèvecoeur's loyalist leanings. Still, perhaps he comments on the way the resilience he sees in the colonists will allow them to overcome a tyrant. What is actually more interesting here than the bees and the king-bird themselves is James's role in the ordeal. He, and thus American humanity, takes on the guise of a divine hand (the same hand that in a single motion ended the futures of several egg-encased chicks). It is his American hand—a form of nationalistically charged divine intervention—that shoots the king-bird, that removes 171 bees from its craw, that places them in the sun, and that is responsible for the reanimation of 54 of them. Of this turn of events D.H. Lawrence, in his *Studies in Classic American Literature*, wrote "I don't care whether it's true or not. I like the picture, and see in it a parable of the American resurrection."[31] So, if Crèvecoeur does, in this fable, comment upon the colonists' need to foster and leverage communal power in the interests of individual freedom, it would seem he gestures toward not so much a form of divine intervention as an understanding of how valuable lessons are taught every day in the tiny conflicts and injustices that dot the natural world. Ultimately, the lesson of Crèvecoeur's odd, ambiguous allegory takes a backseat to his choice of instructional material. Here, as elsewhere, he employs the smaller players of the natural drama—insects, birds, snakes, plants—to instruct humans on human matters. As Stone writes, "Here man, perceiving nature as both predatory and beneficent, intervenes like God in the natural order, plays the dual role of judge and midwife, and helps to tip the balance, making nature in this case more kind than cruel."[32]

When considered in association with the fable of the king-bird and the bees, Crèvecoeur appears to reinforce the conclusion that his allegory illustrates, above all, the need for the human colonists—in their struggle against King George III—to demonstrate the kind of vigilance that the bees, to their peril, fail to sustain. Along with the bees' perceived loss of diligence comes the loss of their strength in numbers. That James steps in to kill the king-bird and extract the swallowed bees is presented merely as a way to satisfy his own distaste for the victimization of the bees, whom he much admires. It would seem Crèvecoeur views humanity as both charge and master of the natural world. Humans must cultivate the land to survive, but humanity also has a responsibility to punish those animals seemingly

---

30 Osborne, "American Antipathy and the Cruelties of Citizenship in Crèvecoeur's *Letters from an American Farmer*," 535.

31 Lawrence, *Studies in Classic American Literature*, 36.

32 Stone, Introduction in *Letters from an American Farmer*, 21.

at odds with accepted human morality. After all, a reluctant James decides to kill the king-birds only after their greed threatens the balance of species on his farm.

Crèvecoeur promotes the idea that a type of natural moral balance should exist in the animal world. He portrays James as the American everyman who is somehow ordained to dispense punishment to unwary animals. This trope occurs yet again in an insect allegory found in his *Sketches of Eighteenth-Century America*. Composed of a series of essays that were excluded from *Letters* for a reason unknown, *Sketches* reinforces and deepens several elements common to Crèvecoeur's portrayal of the natural world, including the animal fable. In the chapter entitled, "Thoughts of an American Farmer on Various Rural Subjects," he writes, "If bountiful Nature is kind to us on the one hand, on the other she wills that we shall purchase her kindness not only with sweats and labour but with vigilance and care" (297). Here Crèvecoeur makes explicit the idea of natural balance hinted at in *Letters*, and it is this idea that he further examines at a key point in *Sketches*. Specifically, he follows his idea about natural balance with an object lesson in how humanity should not only intercede into natural affairs as a figure of deliverance, but also as a dispenser of punishment. At this late point in the chapter, James, while walking in the forest, encounters a tree bearing a beehive. Always interested in the well-being of the bees, a creature that occupies a rich figurative territory in Crèvecoeur's writing, James notes that the hive is being raided by a group of ants, who, travelling to and fro in thick columns, are stealing the honey from the bees. James relates, "I killed a few which I smelled. I found them all replete with honey. I therefore concluded that these were a set of thieves living on the industrious labours of the others" (295). This entomological injustice so bothers James that he returns the next day and exacts punishment against the ants:

> Next day, accompanied by my little boy, I brought a kettle, kindled a fire, boiled some water, and scalded the whole host as it ascended and descended. I did not leave one stirring. The lad asked me a great many questions respecting what I was doing, and the answers I made him afforded me the means of conveying to his mind the first moral ideas I had as yet given him. On my return home I composed a little fable on the subject, which I made him learn by heart. God grant that this trifling incident may serve as the basis of a future moral education. (295)

This brief account represents an intriguing confluence of elements found in Crèvecoeur's animal anecdotes in *Letters* and his general discussions of the American natural world. As with the famed passage about the king-bird and the bees, Crèvecoeur's James discerns a moral lesson to be had in the abuse the bees suffer from another non-human denizen of the environment. Just as James decides to kill the king-bird as punishment for its greedy consumption of bees, so too does he exact a similar execution of the ants engaged in stealing another beehive's honey. Predictably, given the moralistic lens through which he (Crèvecoeur?) views the natural world, James interprets the ants' taking of the honey as a horrendous theft, a crime instead of a natural, amoral occurrence. Therefore,

Crèvecoeur's portrayal of the human role as a godlike dispenser of violent justice against the trespasses seen in the natural world remains consistent. An interesting change in this passage, however, is that James takes pains to utilize this natural occurrence as a direct lesson to his son. In fact, James goes to great trouble in service of this idea. He expressly returns the next day with his son, sets up the labor-intensive apparatus of punishment (boiling water in the woods), repeatedly scalds the ants at the tree in question, and engages in a lengthy discourse with his son whom James reports understandably "asked me a great many questions respecting what I was doing." James then crafts a "fable," based upon the event and insists that his son memorize the tale to ensure that the experience would "serve as the basis of a future moral education." As Jeff Osborne has recognized in other of Crèvecoeur's scenes of nature's activities, passages such as this one invoke an intrinsic cruelty in Crèvecoeur's scenes of nature, but that cruelty is itself part of Crèvecoeur's larger allegorical project. "Crèvecoeur's notion of a cruel Nature," writes Osborne, "should be understood rhetorically: that is, as a discursive component of Crèvecoeur's use of James as ironic and parodic persona; as a means by which Crèvecoeur teases out how an unconscious antipathy informs American expressions of sympathy; and, ultimately, as the fulcrum of Crèvecoeur's critique of liberal contract theory."[33] Thus, at many levels, Crèvecoeur combines the idea of gleaning human moral lessons from the perceived injustices found in nature, with the idea that such lessons, if they are to have lasting and transmittable moral value, must take on a literary form. Such fables, as he terms them, transform the natural world into a showcase for metaphor-laden lessons on human morality.

The self-referentiality housed in this anecdote sheds much light on his use of the natural world across his works. Of note is Crèvecoeur's seeming obsession with animals of the order Hymenoptera (which includes bees and ants) as stand-ins for human communities. His revolutionary bees suffer at the beak of the king-bird only when they break ranks, when they abandon the military discipline required to leverage their strength in numbers. The thieving ants likewise victimize the bees whose honey they steal. That the ants themselves are another type of swarming, colonial creature seems of some importance to the parable. Ants are, after all, likely perceived by James as something of a lesser insect because they do not fly and cannot generate a commodity and service useful to humanity (namely, honey and pollination). So they become in Crèvecoeur's moralistic American wilderness a metaphor for a community given over to crime either from a natural tendency toward shiftlessness or at least a lack of productivity. Whether James's method of killing the ants is cruel or just the most effective (scalding them with boiling water invokes images of Hell) Crèvecoeur once again positions James as a vengeful god quick to pronounce judgment upon any perceived crime among the animals.

---

33 Osborne, "American Antipathy and the Cruelties of Citizenship in Crèvecoeur's *Letters from an American Farmer*," 537–8.

Conclusion

Crèvecoeur and Bartram concern their hybridized nature writing with the imagined construction of human communities in a natural environment filled with moral lessons. Bartram constructs such communities throughout his writing, such as when traversing a river, in *Travels* he describes not only a diverse community of wildlife but one that carries with it a lesson on human conduct:

> It is amazing and almost incredible, what troops and bands of fish and other watery inhabitants are now in sight, all peaceable; and in what variety of gay colours and forms … continually peopled with prodigious numbers and variety of fish and other animals; as the alligator and the manate or sea cow. (198–9)

Here he describes a community in general terms and makes special note of their cohabitation and diversity of species. The passage leaves the reader wondering if Bartram realizes the anthropomorphic irony behind describing the river as "peopled" with fish. His southeastern American wilderness most certainly qualifies as "peopled." After all, the wilds of *Travels* harbor a remarkable diversity of life, including the towns of the American Indians who, through their relationship with the natural world, helped Bartram mold *Travels* into a book that urges readers to better understand the human place within a much larger and ultimately inescapable natural community. Bartram's recognition of such interactions is a testament to his forward thinking and his ability to view the southern wilds as an assemblage of dynamic communities. Bryan Moore contends that Bartram was "ahead of his time in his positive attitude toward the simple living of the American Indians, whom he links more than once with wild animals—an affirmative comparison in Bartram's case."[34] While the lives of the Native Americans Bartram encountered were probably anything but "simple," his admiration for their way of life does seem genuine. As one of America's first romantic writers, Bartram, could never resist the emotional appeal of personifying the natural world around him. Indeed, Thomas Hallock writes that Bartram's "romanticism ... reflects an awareness of incipient change, articulated by a negotiation across modes of written expression and against the context of an expanding nation. The imagined wilderness of the text, however far its creator might wander, remains in the frame of a Euroamerican historical geography."[35]

Similar to Bartram's clouds of mayflies and schools of fish, Crèvecoeur's James often employs swarms of bees and ants as anthropomorphic placeholders for human communities. Interestingly, when Crèvecoeur shifts his discussion from the community to the individual, his insect metaphors tend to give way to plants. These instances provide some of the book's more thought-provoking passages. For example, writing in the famous chapter titled, "What Is an American?" he states that America is an "asylum" for downtrodden Europeans:

---

[34] Bryan L. Moore, *Ecology and Literature*, 106.

[35] Hallock, *From the Fallen Tree*, 164.

> Urged by a variety of motives, here they came. Everything has tended to regenerate them: new laws, a new mode of living, a new social system; here they are become men: in Europe they were as so many useless plants, wanting vegetative mould and refreshing showers; they withered, and were mowed down by want, hunger, and war; but now by the power of transplantation, like all other plants they have taken root and flourished! (69)

Crèvecoeur continues this metaphor, so appropriate for his farmer narrator, writing, "Men are like plants; the goodness and flavor of the fruit proceeds from the peculiar soil and exposition in which they grow. We are nothing but what we derive from the air we breathe, the climate we inhabit, the government we obey, the system of religion we profess, and the nature of our employment" (71). Crèvecoeur's clever botanical calculus equates the American individual both to the literal environment of North America (an environment that directly nourished and housed and bound the individual) and to the sociopolitical environment. Where these two realms, the natural and the artificial, join most elegantly is in what Crèvecoeur refers to as "the nature of our employment." For the majority of Americans, including the narrating James, that employment was farming, a profession that required an intimate knowledge of the environment and also a sophisticated method of subversion, of artificially directing nature's productions to the benefit of humanity. That Crèvecoeur figuratively makes plants of farmers, indicates a recognition of the profound importance of the natural world upon every aspect of American life.

When taken in association with his portrayal of the natural world, we can derive at least one answer to Crèvecoeur's question of "What is an American?" Americans, it would seem, are a people so intimately connected to the natural world that they recognize the fact that they are themselves a product of their own agricultural dependence on the land; they are the fruit of their own tilled soil, a crop whose entire life and destiny is inextricably connected not only to the diverted fecundity of the land but also to the constructs of government, religion, and professional identity. In short, both Bartram and Crèvecoeur argue in their disparate ways that Americans are intensely aware of how the natural world influences their fledgling and amalgamated culture, and—as we will see in the next chapter—this idea becomes further magnified in later, overtly fictional, works that place the American experience and its ambiguous identity squarely upon the land itself.

Chapter 2
Landscapes of Captivity

> What we wish to escape to is not "nature" but an alluring conception of it, and this conception is necessarily a product of a people's experience and history—their culture. Paradoxical as it may sound, "escape to nature" is a cultural undertaking, a covered-up attempt to "escape from nature."
>
> —Yi-Fu Tuan, *Escapism* (1998)[1]

Following the Revolutionary War, the development of a distinct body of American fiction was slow in coming, and predictably, any sophisticated portrayal of the natural world within these works was equally tentative. Yet in the decades preceding the publication of James Fenimore Cooper's 1823 novel *The Pioneers*, a handful of fiction writers did attempt to make the American natural world more than simply a backdrop to their tales. For those few authors who did afford the natural world a more prominent place in their stories, most presented a one-dimensional portrait of a treacherous land beyond the settlements. Some writers, however, succeeded in offering a depiction of the land that shed its typical passivity. These writers presented a more dynamic natural world that was, to varying degrees, intimately involved in the plot and that factored into the joys and fears of the characters. Despite their authors' tendency to ultimately fall prey to fancifully pastoral representations, these instances in which the natural world takes on a more pronounced place in the story are well worth examining. Such works demonstrate the gradual formation of an American literary conception of the natural world, while also highlighting the significance and originality of Cooper's later experiment of writing a direct, complex engagement with a fictionalized, but realistic, American environment.

In this chapter I examine four early American texts that offer a sample of how the natural world was imagined in the fiction of the new republic. These works include the pseudonymously published Panther Captivity narrative (1787), Ann Eliza Bleecker's *The History of Maria Kittle* (1790–1791), Gilbert Imlay's *The Emigrants* (1793), and Charles Brockden Brown's *Edgar Huntly* (1799). It would be overly generous to designate these early narratives as solidly environmental texts. They offer instead brief glimpses of the possibility of U.S. environmental fiction, not so much blazing a trail for future generations as indicating that such a path might one day exist.[2] These proto-environmental texts share another

---

1 Tuan, *Escapism*, 19.

2 The potential domestic European American readership was decidedly small: only about 5 million European Americans inhabited the area east of the Mississippi River (a number that would rise to 76 million by 1900); see Opie, *Nature's Nation*, 84. It was, however, a highly literate population, one in which the vast majority of adult men and women could read (even if not all could write), see Hall, *Cultures of Print*, 172.

significant trait: each are either outright Indian captivity narratives or, as is the case in the two novels, incorporate important moments of Indian captivity, a fact that tells us that for these writers, and probably for their intended readers, the American natural world—whether or not it was directly addressed as such—was characterized as a place that held the potential for intense violence rising from racial oppression.

Thus the American natural world, as portrayed in the final decades of the eighteenth century, is a land encumbered with a weighty set of significations that at times seem to contradict one another. In these and other texts of the period, the natural environment, especially that which might be defined as wilderness, represents a great many things depending on what you expect to find. It was a realm of picturesque beauty that likewise possessed the potential for horrifying violence and misery. It was a land that offered a new beginning and perhaps freedom or agency to those who could not find such privilege in the more developed areas back east, or over the ocean. But as we will see, it was also a landscape that held the very real threats of captivity and murder at the hands of American Indians, a people whose lives and very way of life itself were under severe assault by the genocidal policies of the Europeans. Still, for most, in the nation's first years the land was a cornucopia of social stability and individual gain. Historian John Opie writes, "The new Americans possessed environmental and social advantages that would be envied today by a Third World nation ... The average standard of living was much higher than in Britain or Europe."[3]

Yet the specter of Europe still loomed over the war-weary nation. According to Larry Kutchen, some critics have misread the picturesque literature of late eighteenth-century America. Because they approach these works with "romanticist notions of nature and a Turnerian definition of national development—the dialectic between self and wilderness—they devote too little attention to the picturesque's imperialist founding and dimensions."[4] Kutchen further argues that

> the picturesque emerges in British America not as a mediation of the reality of wilderness, but rather of the real and present danger of revolution ... In their adaptations, 'wilderness' in North America was the property not only of the British empire but also of a British imperial aesthetic, through which the chaos of revolution could be represented and controlled through its displacement in landscape.[5]

Kutchen's warning to read these texts without anachronistic critical baggage is well taken. Also helpful is his reminder that the trauma of the recent revolution was still firmly on the minds of these authors; indeed some of the writers his article

---

[3] Opie, *Nature's Nation*, 119.

[4] Kutchen, "The 'Vulgar Thread of the Canvas': Revolution and the Picturesque in Ann Eliza Bleecker, Crevecoeur, and Charles Brockden Brown," 395.

[5] Kutchen, "The 'Vulgar Thread of the Canvas': Revolution and the Picturesque in Ann Eliza Bleecker, Crevecoeur, and Charles Brockden Brown," 396–7.

addresses, such as Bleecker and Crèvecoeur, had suffered great personal loss during the war. With that said, Kutchen seems to understate the reality that these writers were, as the selection of works in this chapter makes clear, well aware of another danger in the wilderness, the Native Americans, a group of people who had already been deeply wronged by imperial crimes, and who unlike the British armed forces, did not simply pack up and leave at the end of the war.

So regardless of any anxieties born of the revolution, North America was still a land wrenched unjustly from entire nations of Native Americans, who in their desperation resisted the transgressions of the Europeans by raiding settlements, destroying homes, and tormenting, killing, assimilating, or ransoming captives. Readers thrilled at such tales of Indian captivity, and texts in this genre represent some of the first truly popular works of American literature. One of the goals of this chapter is to investigate how, in these representative texts of early republic fiction, writers worked to resolve this paradoxical portrayal of the natural world. If, as some scholars have persuasively argued, the Indian captivity narrative is ultimately a tale of separation, transformation, and return, then how does the environment function in this process?[6] How does it become complicit in the initiation of European Americans in what was to them a land at once beautiful and hostile?

The Feminized Land of the Panther Captivity

In 1787 a brief captivity narrative presented one of the first entirely fictionalized portrayals of the American natural world. Its plot is summarized by the narrative's long title, *A Surprising Account of the Discovery of a Lady Who Was Taken by the Indians in 1777, and After Making Her Escape, She Retired to a Lonely Cave, Where She Lived Nine Years*. Critics tend to abbreviate the title to just "The Panther Captivity Narrative," after the pseudonym of its supposed author, one "Abraham Panther." According to Kathryn Zabelle Derounian-Stodola, between 1787 and 1814 this sensational narrative saw print (usually in pamphlet form) over twenty times, making it something of an early American best seller. The popularity of this narrative probably came in part from its dramatic plot. This brief epistolary account, presented as a single letter to a friend, relates how the narrator and a companion set out "to penetrate the Western wilderness as far as prudence and safety would permit."[7] As Derounian-Stodola contends, the Panther Captivity was most likely influenced by John Filson's 1784 *Discovery ... of Kentucke*, which contained as an appendix the sensational and extremely popular account of Daniel Boone's adventures.[8]

---

6 See VanDerBeets, *The Indian Captivity Narrative*, 44–50, and Derounian-Stodola and Levernier, *The Indian Captivity Narrative, 1550–1900*, 40.

7 Panther, *A Surprising Account ...*, 86. Subsequent references appear parenthetically in the text.

8 Derounian-Stodola, *Women's Indian Captivity Narratives*, 83.

Both Filson's Boone and the Panther Captivity present a western wilderness that is, at least on the surface, possessed of an Edenic beauty that to an extent overshadows the threats it harbors in the form of wild animals, cruel Indian attacks, and the risk of death from exposure, each of which also form a significant part of such accounts. For example, after traveling for 14 days, the men of the Panther Captivity find themselves deep in the wilderness, where they ascend "a high hill, at the foot of which ran a beautiful stream, which passing through a small plain, after a few windings lost itself in a thicket, and observing the agreeable picturesque prospect which presented itself on all sides" (86). From this description we cannot help but imagine the men traveling through a harmless woodland paradise, a realm of predictably and well-worn romantic pastoral beauty. However, once immersed within this Edenic landscape, the men are "surprised at the sound of a voice which seemed at no great distance" (86). The voice sings "a mournful song," (86) and while the men must have known it belonged to a woman before they followed the sound to its source, they seem oblivious to how the wilderness itself, once filled with that voice, becomes immediately feminized and sexualized in its portrayal. Following the voice to its owner, they walk along

> a small foot path which we followed, and arriving at the top of the hill, passed round a large rock, then through a thicket of bushes at the end of which was a large opening; upon our arrival here, to our inexpressible amazement, we beheld a most beautiful young LADY sitting near the mouth of a cave. (86–7).

This sexualized, indeed figuratively pubic, wilderness makes available a range of interpretive possibilities, but for those so accustomed to reading about the wilderness as Mother Nature, the strongly feminized landscape might indeed go underappreciated simply because writers tended to present the wilderness as a woman, a practice equally prevalent today. Yet the degree to which this setting becomes akin to the genitalia of a woman reminds us that these men left with the express intent "to penetrate the Western wilderness" (86). Nonetheless, there is an ambiguity about the narrative, one that contributes to its wide appeal. Annette Kolodny writes,

> What makes the Panther Captivity so unusual is that it adhered to the essential male fantasy of woodland intimacy while, at the same time, it offered a positive image of the white woman's capacity to survive and plant gardens in that same wilderness. In so doing, it catered, in symbolic terms, to opposing and contradictory dreams.[9]

What makes this feminized landscape so potent in this fictionalized captivity narrative is not so much its pronounced sexual imagery but the woman herself, whom the men treat initially as they might an unknown wild animal, and who, in

---

[9] Kolodny, *The Land Before Her: Fantasy and Experience of the American Frontiers, 1630–1860*, 66–7.

recounting her tale to the men, partially justifies their fear of this strange "LADY" (her class further complicates the scene) in the woods. With her singing covering the sound of their approach, the woman's dog first views the men and barks at them. Turning in surprise, the woman faints upon seeing the men. With this swoon the gender dynamic begins to realign and the woman begins to move—in the estimation of Abraham Panther and his companion—from the status as an odd feature of the forest to a female in need of male rescue. It happens in a moment, as if the mere presence of white men in the feminized wood tames her and reclaims her for the patriarchy: "she started up and seeing us, gave a scream and swooned away. We ran to her assistance and having lifted her up, she soon recovered; and looking wildly at us exclaimed, Heavens! Where am I! and who, and from whence are you?" (87).

The woman, who had not seen another person for many years, tells the men of how she has eloped with a man disapproved of by her father, and of how as they traveled through the wilderness they endured an Indian attack that ended with the torture and death of her husband. She escapes this nightmare only to be captured by "a man of a gigantic figure" who lives in a cave deep in the wilderness (88). One day the giant indicates that she "must either accept of his bed or expect death for my obstinacy" (89). Horrified by the threat, she waits until the giant is sleeping, then kills him with his own hatchet, and, in a grotesque but practical act, cuts off his head and hacks his large corpse into quarters, dragging it from the cave to a shallow grave in the woods ("covering him with leaves and bushes" [89]). Free of her monstrous oppressor but still lost in the woods, the woman relates that she took up residence in the cave, where with a dog as her only companion she lived in relative comfort for nine years, sustaining herself by gathering nuts, berries, and apples from the forest while also cultivating a small crop of Indian corn.

Reviewing some of the critical possibilities of this popular captivity narrative, Derounian-Stodola writes that the story "lends itself to a host of interpretations including mythic/historical, Freudian, and feminist."[10] While Derounian-Stodola's various critical approaches focus predominantly upon the plight of the female protagonist, a closer look reveals how the woman and the wilderness collaborate in latent terms to represent a multilayered threat to patriarchal control. Derounian-Stodola argues that the story forwards "the dilemma facing America as an independent republic with two models to chose from: the Europeans' world ... and the American Indians' world (signified by the wilderness locale),"[11] and while Derounian-Stodola's interpretation is meant as just one of several possible avenues of investigation, this one seems to overlook the way that the problem of masculine domination and violence conspire—whether the men are European or American Indian—to drive the woman into the wilderness to make for herself a lonely but peaceful and decidedly feminine sanctuary. After all, the woman flees her father's control only to fall into the hands of murderous Indians. Fleeing those

10 Derounian-Stodola, *Women's Indian Captivity Narratives*, 84.

11 Derounian-Stodola, *Women's Indian Captivity Narratives*, 84.

men (who kill her husband), she finds herself in an equally dangerous situation as the prisoner of the giant who lives in the cave. Only after killing the giant does she escape patriarchal control for the first time in her life. Nine years later, the arrival of Abraham Panther and his traveling partner instantly places her back under male authority, and even though by the end of the story she returns to visit her father's deathbed and inherit his fortune, she may never enjoy the liberty she knew during those years in the wilderness. So, for the woman reader this narrative presents not so much two models for European American nation building as a model for how to escape the European American project altogether. The wilderness of this narrative becomes a space representing the absence of male domination regardless of the race of the men in question. But as Amy Pratt points out in her examination of the economic and identity issues housed within the narrative,

> Like the cave itself, which opens out into a series of interior apartments, the experience recounted here suggests an interior self with unknown depths and unexplored possibilities ... The psychological depth that offers the promise of expansion to a "natural" interior self is symbolically evoked and questioned in "The Panther Captivity" both by the darkness implied by the figure of the cave and the silence that takes the place of any details about the long time there.[12]

The Panther Captivity highlights this anti-patriarchal function of the wilderness in its presentation of the woman character's incremental—if sometimes unwanted—liberation from a sequence of men, a sequence that begins after she and her new husband enter the deep woods. Avoiding the wrath of her father—who detests her suitor so much that he "hired several men to search the country in pursuit" of them—the young couple decide to take up residence "further back into the country" (88). Once in the wilderness, where she is beyond the reach of her father and his henchmen, the woman also immediately loses her husband, who is cruelly murdered during an Indian attack. Fleeing the Indians, the woman soon becomes lost in the unexplored wilderness, and eventually she comes across the man in the cave. This man, who is presented as an Indian and a giant, represents the patriarchal dominance the woman has suffered her entire life. To be free of the giant's sexual advances, and more figuratively the violence and control of men in general, she must get rid of the giant man and take possession of the female potential in the wilderness, here embodied as a vaginal space by the cave in which the giant lives. The woman is finally free of the patriarchy only after she slays the giant and moves into the cave. Derounian-Stodola recognizes this last part of the woman's process of shedding patriarchal control when she writes, "the independent woman brandishing an axe and a knife becomes every red-blooded (white) male's nightmare: she is woman as castrator. In the right circumstances,

---

[12] Pratt, "The Pleasure of Being Lost: 'The Panther Captivity' and the Metaphysics of Commerce," n.p.

this lady—indeed, any woman—can become defeminized and thus no longer under male control."[13]

This psychoanalytical trope of the figuratively castrating woman emerges again and again, both in fictionalized Indian captivity narratives and in those that are more or less based in fact. In explaining, for example, their compelling psychoanalytic reading of Indian captivity narratives, Derounian-Stodola and James Levernier argue that "women ... who survived Indian captivity without their husbands, are said to awaken subconscious male fears of impotence and castration through their ability to live through the experience without male intervention."[14] As their goal is to reveal this trend as one among a rich collection of other critical approaches to this highly popular American literary form, these critics can be excused for not doing more to connect these feminist strains with the tale's setting, which is invariably a hostile wilderness. However, when investigated, it becomes apparent that these narratives' ability to tap into those "subconscious male fears" has much to do with the ways such stories relocate the woman from the European American settlements to the unsettled and unsettling wilds, where white men find themselves no longer in control of the land or their women, and where they must compete with Native Americans, wild animals, the elements, and the ruggedness of the terrain. Christopher Castiglia, in discussing Catherine Maria Sedgwick's novel *Hope Leslie*, writes, "captivity narratives offer women the metaphors necessary to describe their cultural imprisonment and to imagine jailbreaks into lives, if not of freedom, then of re-visionary adventure."[15] A woman in the wilderness may not be safe, but she is—perhaps for the first time in her life—beyond the convenient grasp of the European American patriarchy. If the protagonist also evades Indian captors then she enjoys (whether conscious of it or not) a level of freedom and independence all too rare for women of the day. Such freedom can only be found in the wilderness, making this setting the engine by which women readers could plausibly imagine social liberation.[16]

---

13 Derounian-Stodola, *Women's Indian Captivity Narratives*, 84.

14 Derounian-Stodola and Levernier, *The Indian Captivity Narrative, 1550–1900*, 44.

15 Castiglia, *Bound and Determined*, 164.

16 Kolodny argues that the Panther Captivity's popularity resides in the way it speaks to an alternate female land-use fantasy; in *The Land Before Her: Fantasy and Experience of the American Frontiers, 1630–1860*, she writes,

> it seems that women, at least, held to the dream of creating gardens in the wilderness. And this, perhaps, explains the appeal of the Panther Captivity for that generation of readers who grew to maturity between its first appearance in 1787 and its last known printing in 1814. For them, as they watched the selling off of public lands in the west, its odd pastiche of generic sources delineated what all the Boone narratives evaded: that is, the symbolic terms of the then competing fantasy possibilities—in one of which women's fantasies found their first palpable expression. (67)

Wilderness as Sentimental Mirror in *The History of Maria Kittle*

One of the most intriguing traits of the Panther Captivity is how it seems intended to appeal to female readers while also presenting a tale of adventure in the American wilderness. This literary trait occurs in other captivity narratives, particularly those that were written by women. Take for example Ann Eliza Bleecker's Indian captivity narrative, *The History of Maria Kittle*, a story ostensibly based upon the experiences of Kittle during the French and Indian Wars. If we are to believe Bleecker's opening remarks, the story was apparently told to her by a dying woman named Mrs. C— V—, who heard it years before from Kittle herself. In 1779 Bleecker wrote an account of Kittle's story as a letter to a Miss Susan Ten Eyck, but Bleecker died in 1783, having never published any of her work. Bleecker's daughter, Margaretta, ended the narrative's long road to publication when she serialized it in two installments in *New-York Magazine* in 1790 and 1791.[17]

Reportedly based upon a true story, Bleecker's tale is nevertheless heavily embellished with lush description, copious dialog, and moments of unknowable narrative interiority. These combine to make a story that is clearly fictionalized. "We should remind ourselves," E.W. Pitcher writes,

> only the bare bones of the tale could have been related to Bleecker by Maria Kittle's sympathetic kinswoman. Bleecker embellished the story of Kittle's captivity to create a sentimental and gothic captivity tale ... the dialogue must everywhere have been Bleecker's invention, as is most of the emotion-charged description. When a rhetorical effect is required or a conventional trope demands it, some details are manipulated to suit the occasion.[18]

In fact, as Derounian-Stodola and Levernier have recognized, Bleecker's

> novel is largely fictional and draws heavily upon the conventions of the sentimental novel so popular in England and America during the late eighteenth century ... In the hands of Ann Eliza Bleecker, the trappings of the Indian captivity—the wilderness journeys, the miraculous escapes, the merciless enemy—are all transformed and converted into those of the sentimental novel.[19]

This assessment certainly rings true, but there is more to the novel than just this layer of sentimentality somewhat clumsily draped over an Indian captivity narrative, and that is the way in which this veil of sentimentality, this very tendency to excess emotion, is reflected in the environment of the tale.

---

17 Harris, "Ann Eliza Bleecker, 1752–1783," 2.

18 Pitcher, "The 'Hapless Babes' of the Frontier: Ovid, 'The History of Maria Kittle,' and *The Last of the Mohicans*," n.p.

19 Derounian-Stodola and Levernier, *The Indian Captivity Narrative, 1550–1900*, 186–7.

While *The History of Maria Kittle* does not contain anything resembling a progressive feminist message like that found in a few later works of frontier fiction (compare Gilbert Imlay's *The Emigrants*, discussed later), it does, however, tell the story of a remarkably resilient and intelligent woman who endures the horror of an Indian raiding party that kills several members of her family and takes her captive for many weeks. At the story's outset Maria Kittle, her husband, and her family feel a great affinity and appreciation of the natural world in which they have made a home, and Bleecker offers the usual pastoral postcard descriptions. For instance while out hunting, Mr. Kittle tells his brother, "what a profusion of sweets does Nature exhale to please her intelligent creatures! I feel my heart expand with love and gratitude to heaven every moment, nor can I ever be grateful enough."[20] Such a statement satisfies most of the usual pastoral requirements. Appropriately awed by the transcendent beauty of the natural world, Mr. Kittle thanks God for providing such a paradise in which to live, but Bleecker almost immediately punctures this peaceful scene by having Peter, Mr. Kittle's brother, experience a premonition that something terrible is about to happen. He tells his brother, "an unusual horror chilled my blood; I felt as if the damps of death prest already round my soul" (7). Just a few hours after his vision, Peter dies, shot dead in the forest by the same Indian raiding party that will that night kidnap Maria and kill their infant children. After dispatching two of the attacking Indians, Mr. Kittle runs home, and as becomes the case with several later texts, with the horror that accompanies a murder comes a change in the character of the natural world itself. With the arrival of bloodshed the pastoral playground disappears, and as Bleecker writes,

> When our souls are gloomy, they seem to cast a shade over the objects that surround us, and make nature correspondent to our feelings: so Mr. KITTLE thought the night fell with a deeper gloom than usual. The soft notes of evening birds seemed to be the responses of savage yells. The echo of his tread, which he never before regarded, now rung dismally hollow in his ears. Even the rustling of the winds through the leaves seemed attended with a solemnity that chilled him with cold tremors. (8)

This idea actually works counter to the pastoral ideal. In this view the character of the land is not set entirely by a divine plan; its tone is in some ways dependent upon the moods of those who enter the forest. If they bring darkness with them, then the land darkens around them. Bleecker's landscapes in this narrative are then mirrors of human feelings, meaning that the natural world is connected in an intimate way with the very minds of those who enter it, and who suffer the horrors brought there by others. Those horrors were compounded by the isolation experienced by frontier farmers, and some form of this isolation exists in each of text discussed in this chapter. John Opie reminds us that

---

[20] Bleecker, *The History of Maria Kittle*, 7. Subsequent references appear parenthetically in the text.

> The frontier farm family labored through its days and nights in intense isolation on a lonely separate site, distant from neighbors ... frontier accounts tell of the madness of wives and the violence of husbands. The strongest sense of security may have been the farmer's psychological bond with his tract of land.[21]

Indian captivity narratives such as Bleecker's almost always incorporate this isolation as part of the horror of the experience, and I argue they also include this sense of a psychological bond between humans and the land. As Opie argues, the environment becomes not only a site of home and sustenance but also a site of a psychological link, one in which the land occupies the human residents' minds almost as much as the settlers occupy the physical land.

As Bleecker's narrative abandons Mr. Kittle to follow Maria through her trials as a captive, this idea of a forest that echoes the moods of its inhabitants continues. At one point, Maria—understandably disconsolate at the murders of her child and the torments of captivity—views her surroundings with dread:

> Sad and benighted they wandered through almost impenetrable swamps, forded the broad stream of *Tomhanick* and the rapid river of *Hosack*; they pushed through deserted settlements, where the yelling of solitary dogs increased the solemnity of midnight ... Maria, overcome by sorrow and fatigue, immediately sunk helpless at the foot of a tree, while the savages ... kindled a fire, and prepared their meal. (14)

Bleecker highlights this strangely changeable wilderness by having Maria awaken to a beautiful scene that is nonetheless spoiled by her plight as a captive, making for a realm of intense beauty that seems to alternately mirror and mock her suffering:

> she was the first that awaked as the sun began to exhale the crystal globules of morning, when half rising, and reclining on her elbow, she surveyed the lovely landscape around her with a deep sigh; they were on an eminence that commanded an unlimited prospect of the country every way. The birds were cheerful; the deer bounded fearless over the hills; the meadows blushed with the enamel of FLORA but grief had saddened every object in her sight; the whole creation seemed a dark blank to the fair mourner. (17)

Just as Mr. Kittle observed how his fearful mind made an otherwise pastoral landscape into a gloomy realm, so too does the tormented Maria hyperbolically romanticize her surroundings while simultaneously filling them with dread. Still, despite her misfortune, Maria does not perceive her natural world (for her overly romanticized view is doubtless hers alone) itself as overly hostile. She finds, even in her depressed state, some scenes inescapably beautiful and in so doing derives a fleeting comfort from them. For instance, when her captors stop by the bank of a stream she finds herself envying the fish:

---

[21] Opie, *Nature's Nation*, 59.

> Observing the spotted trout, and other fish, to dart sportively across the water, she could not help exclaiming, "Happy! happy animals! you have not the fatal gift of reason to embitter your pleasures; you cannot anticipate your difficulties by apprehension, or prolong them by recollection; incapable of offending your Creator, the blessings of your existence are secured to you: Alas! I envy the meanest among ye!" (19)

That Kittle would wish herself transformed into a fish speaks to the terror and misery of her captive condition, and it also demonstrates how the idea of the natural world, here reduced to the emblem of a spotted trout, becomes a vehicle for escape from the painful moments of the human condition. She calls the trout "happy" in their simple lives, free of "the fatal gift of reason" and unable to become mired in the gloom of "apprehension" and painful "recollection." For Kittle, the ignorant lives of the animals serve as a contrast to her own existence, in which her captivity paradoxically highlights her humanity. She is a woman who has become a caged creature.

It is telling that she, like her predecessors in the Panther Captivity and in so many other captivity narratives, finds solace in the otherwise antagonistic wilderness. The same woods that harbor the Indians and that serve as a form of prison also allow for a type of imaginative escape from the horrors of the physical present and the threat of impending death. As with those previous narratives, it is instructive to keep in mind Castiglia's argument that there is another form of escape in play in these narratives: the escape offered to readers, especially women. Castiglia's contention that captivity narratives speak to their readers' "cultural imprisonment" and provide a venue by which those readers could "imagine jailbreaks" into lives free of patriarchal control seems particularly astute when coupled with the function of the natural world as a magnifier of any such escape.[22] The forests in Bleecker's tale then share in the creation of these metaphorical escapes. The reader cannot transform herself into a spotted trout, but she can certainly imagine herself as the brave Kittle, who is as much a prisoner of the patriarchy as of the Indians who momentarily transferred her from one system of oppression to another.

Bleecker's narrative stands as perhaps one of the best examples of how American women writers of the late eighteenth century increasingly shared their own stories, offering their own visions of the natural world in which those stories were set. At least in terms of their surface appearance these portrayals of the American natural environment were similar to the romanticized pastoral gardens or brutal wildernesses forwarded by their male counterparts. This initial similarity is understandable; women writers had read plenty of works by male, or in the case of the Panther Captivity, ostensibly male, writers. And it was these works from which they undoubtedly took some inspiration. While male writers of the period rarely served as proto-feminists, often making their female characters ingénues who appreciate the natural world on a superficial level, they did still make their female characters part of their wilderness adventures. One example

[22] Castiglia, *Bound and Determined*, 164.

is Gilbert Imlay's *The Emigrants* (1793), an early American novel with a strong environmental slant and an equally strong female character named Caroline. As we will see, while that portrayal is still marked by the tendency to present female characters as props on a forested stage, the book does forward a more progressive and complex view of the relationship between women and the pastoral vision of the natural world so popular in the new republic.

Captivity and the Picturesque in *The Emigrants*

Amplifying the linkages between issues of women's social liberty and the natural environment in the Panther Captivity and *The History of Maria Kittle*, Imlay's epistolary novel, *The Emigrants*, stands as one of the most intriguing, if lesser-known, texts to imagine the American environment. It even incorporates its own Indian captivity narrative, which, like the Panther Captivity, combines the idea of female liberty with a portrayal of the environment that at moments goes beyond the surfaces of a romantic landscape. The extent to which Imlay hoped to create a substantial literary work was almost certainly subsumed by his desire to make a buck in land speculation. *The Emigrants* appeared one year after he had published an ostensibly non-fiction travel account called *A Topographical Description of the Western Territories of North America*. As Imlay had invested in real estate in that very locale, these two books served as promotional tools, with each offering a decidedly romanticized view of the area.[23] In *The Emigrants* Imlay concocts an Ohio River Valley that is at turns a realm of pastoral splendor and frontier excitement (if not outright danger). Set predominantly on what was then the frontier wilderness of the Ohio River Valley, the novel relates the romantic entanglements, general hardships, and, to a lesser degree, wilderness adventures of an English family led by a financially ruined former London merchant named Mr. T—n.

Having fallen upon financial difficulties brought about by the profligate spending of his wife and son, Mr. T—n and family sail to America. Upon arriving in Philadelphia the family begins their trek West, with the plan to "remove into the back settlements of the country, and there in obscurity endeavor to preserve that tranquility, to which the follies and dissipation of Mrs. T—n and his son had made him a stranger."[24] What Mr. T—n lacks in funds he makes up for in well-meaning friends who do their best to ensure the safety of the family as they make their way to Pittsburgh and points beyond. Mr. T—n's beautiful daughter Caroline, a young woman who seems to charm everyone she encounters, soon becomes the central figure of the tale, gaining the quick attention of an upstanding young American named Captain Arl—ton, who, in doing his best to make good on his promise to keep the family safe from the dangers of the American wilderness, quickly

23 Thomas Ruys Smith, *River of Dreams*, 26.

24 Imlay, *The Emigrants*, 5. Subsequent references appear parenthetically in the text.

becomes her suitor. Predictably, Arl—ton's courtship of Caroline takes up much of the novel, but of special note are the ways that the frontier wilderness, often transformed by Imlay into a highly romanticized pastoral paradise, functions as a kind of playing field upon which the two youths—particularly Caroline—test, negotiate, and ultimately fall back into established gender expectations.

Therefore, Imlay's *The Emigrants* provides another instance of what Castiglia contends was a tendency for early women's captivity narratives to allow their female readers to "imagine jailbreaks into lives ... of re-visionary adventure."[25] Unlike the Panther Captivity, the Indian captivity narrative that occurs in *The Emigrants* is not in itself the main vehicle by which this imagined liberation arrives. The actual Indian captivity found in *The Emigrants* comes late in the novel, is quickly resolved without bloodshed, and is almost presented as a narrative afterthought. What instead serves as the vector for imagined "jailbreaks" are those moments, occurring at key points in the novel, when Caroline confronts the awesome spectacle of the wilderness, a landscape that in absorbing her family and its financial problems, also comes to define key aspects of her present happiness and domestic future. Yet the wilderness of Imlay's *Emigrants* is also at moments considered in its more traditionally adversarial light as an evil and treacherous space that simultaneously threatens to swallow up the European American settler and spew forth the likes of wild animals and hostile Native Americans. Well aware of these ways of describing the American wilderness, Imlay overtly addresses them in his narrated introduction to the novel, where he writes that the T—n family was "preparing to traverse a region which they had been taught to believe was inhabited mostly by wild beasts and savages" (5). Once the family begins its journey west, Imlay begins to counter and complicate this preconception, and he does so through Caroline's charmingly naïve pastoral vision of the land. Rather than fear the landscape, she adores it, and her adoration works to transform its character to those around her. For example, a visiting family friend, Mrs. W—, relates in a letter to Mrs. R— (another friend) that

> The fatigue of travelling to this country is merely imaginary, and I am sure if you were to hear Miss Caroline's T—n's description of the mountains and the beautiful landscapes upon the Susquehanna, you would be quite enchanted with the idea of taking such a journey.
>
> I confess, though I had been highly delighted with the romantic and sublime scenes which were continually presented to my view as I came along, still I did not receive a just impression of their various beauties, untill this amiable girl depicted them in such glowing colours as made me think I must have been stupid not to have noticed them. (23)

Imlay, in his various epistolary voices, does not so much counter the established portrait of the American wild as emphasize its ultimate subjectivity and variability

25 Castiglia, *Bound and Determined*, 164.

depending upon whose eyes we see it through and under what conditions they experience it. As Wil Verhoeven claims,

> *The Emigrants* in many respects simply puts into fictional form the ideological concerns of his [Imlay's] early *vade mecum* for the prospective emigrant, *A Topographical Description of the Western Territory of North American* (1792): both give us a taste of a Jacobin version of rural England transplanted to American soil and scenery.[26]

Caroline sees the wild landscape as a series of "picturesque scenes of this romantic country" (51) in which the Susquehanna bears "high impending banks that rudely seem to threaten its limpid stream, as though they were jealous of such attracting beauty" (38). To take full advantage of these scenes, Caroline insists on walking behind the family's wagon a considerable distance over the Allegheny mountains: "I had previously made up my mind upon walking over the mountain, which is about fifty miles across, from the opinion that there must be many interesting views on the way, as I should loose them in case I kept my seat in the waggon" (38–39). Once tired, she takes a break by sitting on a rock covered in green moss, and it is while perched upon this seat, surveying the Arcadian view below, that she first encounters her future husband, Captain Arl—ton. In his effort to catch up with the T—ns he sees her sitting on her rocky perch enjoying the view:

> I first caught sight of Caroline—she was resting upon a large stone on the road side, when I came upon her by surprize,—was leaning with one hand upon her cheek, and held a handkerchief, which I thought had been applied to her eyes, that were glimmering like the rays of the sun through the mist of an April shower. (15)

As Arl—ton's description reveals, he—like his future bride—is given to forming pastoral descriptions. He is, however, less able to understand her love of walking and being alone in the wilderness. He and Caroline strike up a conversation as they walk together, and he relates that "All this time it appeared to me not a little strange, that Caroline should be alone, and walking" (16). She even refuses his assistance in scaling some of the more challenging declivities in their path. Such a refusal, when coupled with her pastoral gaze, suggests that for Caroline, the solitary walk in the woods represents a form of environmental escape from her troubled family and from the controlling hand of the patriarchy.[27]

---

[26] Verhoeven, "'Some corner of a foreign field that is forever England': The Transatlantic Construction of the Anglo-American Landscape," 158. See also Seelye, "The Jacobin Mode in Early American Fiction: Gilbert Imlay's *The Emigrants*," 204–12.

[27] Verhoeven reminds us also that Imlay wishes to present the American frontier as a relatively congenial place; in "'Some corner of a foreign field that is forever England': The Transatlantic Construction of the Anglo-American Landscape," he writes,

Once Arl—ton becomes a fixture of the family's travels he remains perplexed at Caroline's nature strolls, and she must work harder to maintain the Arcadian autonomy that her love for walking behind the wagon in the woods affords her. She is clearly taken with the young captain, and Imlay makes it plain that the two will eventually wed, but she also likes the freedom to explore the beauty of the natural world on her own. Arl—ton tells Caroline, "[I]t gives me much pain to see you walking over these rugged mountains, for surely you must find yourself very much fatigued" to which "[s]he replied no, —it was always a great pleasure for her to walk" and more so than in the "shady groves, or the promenades of London" because

> here is a continual feast for the mind—every rock, every tree, every moss, from their novelty afford subject for contemplation and amusement. Look but at yonder towering hills ... whose summits appear to prop the heavens, and then view the various symbols which their chasms produce, and what a sublime imagery does it afford?—What can more resemble the ruins of a great city? (25)

For Caroline the American wilderness is that most stereotypical of romantic landscapes, a pastoral wonderland that, if not possessed of romantic ruins of past civilization, at the very least invokes a likeness in its chasms and hills. Through Caroline's eyes the landscape that the others of her party recognize as a treacherous wilderness becomes an enchanting garden. The love-struck Arl—ton dares not to interrupt her monolog, relating that "Caroline might have proceeded in this way to eternity, and I should not have interrupted her; for such was my astonishment at the fertility of her imagination, that I heard her with astonishment, and gazed at her with the most ineffable transports!" (25). Here Arl—ton makes sure to remind Caroline that the source of her pastoral description is none other than her considerable imagination, one that transforms a wilderness into a paradise. The great liberating beauty Caroline finds in the wilds of frontier America actually resides within her, the symptom of a long-held sexist portrayal of women as overly given to flights of fantasy.

Caroline is not altogether ignorant of how men criticize women for their imagination as well as for their curiosity in matters perceived as falling outside their prescribed domestic realm. One of the most pronounced examples of how Imlay criticizes this sexist dissemination of knowledge, which is overtly related to

---

> the plot of the novel ... emphasizes the ease with which the emigrant may travel from the eastern states west across the Allegheny Mountains and how an elaborate infrastructure of country roads and waterways will be at his or her disposal once he or she arrives there. With its sentimental interest frequently being put to the service of its geopolitics, it is therefore no coincidence that *The Emigrants* at times reads more like a travelogue than a novel. Thus, the opening letters make much of the heroine's insistence to walk much of the way across the Allegheny Mountains. (158)

So Imlay's portrayal of the landscape, and Caroline's experience of it, is meant to lure people, particularly women, to the prospect of the American frontier.

the comprehension of the workings of the natural world, occurs in a letter Caroline writes to her sister Eliza (who has stayed in London with her husband). As might be expected from one so taken with the beauty of the American interior, Caroline goes to great lengths to describe some aspects of the natural history of that region. It is winter and the young adventurer regales her sister with descriptions of the wintertime activities of the bear ("a torpid animal which sleeps throughout the winter season" [76]), the squirrels (which are "much larger than those of Europe, and ... have most beautiful tales" [76]), and the quail, which in some ways poses an ecological mystery for the inquisitive Caroline, who writes:

> The quail is entirely different from those of Europe. In size they are something larger than a dove, and their plumage very much like the English partridge, from which circumstance they are called by many people in America partridge; but ... they cannot be of that species, as they migrate, and are birds of passage; and what is more remarkable, they are in a degree domestic, for our long lost uncle informs me, when this country was first settled, not a quail was to be seen; but as it became inhabited, they crossed the mountain, and are now in the greatest plenty. (77)

Imlay reveals much about Caroline's character with this letter to her sister, in which she moves beyond the merely picturesque to demonstrate a surprisingly sophisticated curiosity about the actual natural history of the region's wildlife. She pays close attention to how the animals live within this place, how they winter, how they differ from related animals in other areas, and how—as in the case of the quail—they remain enigmatic. Perhaps most boldly underscored of Caroline's recognitions is that the environment and its resident creatures change not only in relation to the season but also in relation to the event of European American settlement. Her quail-related quandary is, after all, how did the quail come to inhabit the place, and why did it do so only after European settlement of the area? "One thing I have not been able to reconcile," she writes, "which is, as they are aboriginal to the continent, and are now found only in cultivated countries, how they existed when America was altogether a wild" (77). She also wonders at their resemblance to those birds found in faraway places and at how they seem to have arrived at the same time as the wagonloads of white settlers entered the area, and she criticizes "the inhuman and unsportsmanlike fowler, [who] will at a single shot destroy a whole brood of them" (77). Caroline's thinking about the dynamics of the natural history of the area, along with the ways that humans alter that natural order, reveals an advanced environmental consciousness on the part of Imlay himself, who seems to invoke through Caroline the idea that human actions have larger environmental consequences.

For Caroline there is another unknown related to the subject of natural history, and that is why women are discouraged from seeking such knowledge in the first place. In concluding her letter to her sister she writes a forward-thinking statement related to the poor state of women's education:

> I do not know, why a person should remain ignorant of a subject that affords so much matter for admiration? and if the education of women have generally been so injudicious, as to prevent their extending their understanding beyond the common limits that custom has prescribed, doubtless it has been the material cause, why illiberal men have estimated our talents at so cheap a rate. For while we have been taught to talk of dress and the things of the day, and which have constituted the extent of our colloquial charms, few women have had strength of mind equal to burst the bands of prejudice. (77–8)

Imlay has Caroline cut to the bone of the problem. Women are not allowed educational opportunities and are thus reduced to surveying only the surfaces of knowledge, a situation that promotes the idea that women are ill suited for understanding the complexities not only of science but also of all forms of inquiry. Imlay's *The Emigrants* has been rightly classed as an early American feminist text. As W.M. Verhoeven and Amanda Gilroy argue in their introduction to the Penguin edition of the novel, the book reveals several feminist traits:

> "*The Emigrants* provides a salutary alternative to the distinct absence of feminist ideals in Imlay's checkered personal life ... especially the rights of women in marriage, which it ties to an anticolonial agenda. The novel exposes marriage in England as a type of cultural captivity for women, and makes a plea for more liberal divorce laws".[28]

And in this passage from Caroline's letter to her sister, Imlay expands his message to include the education of women, linking it directly with the need for a better understanding of the workings of the natural world. Caroline embraces the superficial and largely imaginary pastoral beauty allowed for those of her sex, but she also asks scientific questions and bemoans the lack of formal education women receive. Still, she is no feminist firebrand; she is comfortable venting to her sister, but never makes any feminist demands of Arl—ton, who places Caroline back within the traditional female role of the damsel in distress.

While Caroline's inquisitive intellect seems to make her immune to worry about most of the traditional threats found in the American woods, she does express concern about one common hazard found in the wilderness at the time: Native Americans. When she and Arl—ton meet some locals on the road, she asks them, "[T]here I hope are no Indians in these mountains?" and the residents assure her, "[T]here are not any within two hundred miles of us, and if there were, they should not hurt you, for we would guard and protect you at the hazard of our lives" (26). Thus according to the locals one of the most terrible threats to European American peace of mind in the wilds of North America has ceased to be a threat. Further contributing to Caroline's sense of security is the pledge by the local people to protect her even if it were to threaten their own safety. Yet the potential danger of Indian attack persists as the novel progresses. Just a few letters

---

[28] Verhoeven and Gilroy, Introduction, *The Emigrants*, ix–x.

later, she, Arl—ton, and two friends go out for what Arl—ton refers to as one of "her little excursions in the neighbourhood, in order to view the various prospects and wild beauties with which it abounds" (48). As Arl—ton relates,

> In one of these peregrinations in the course of this day, and in which she was attended by her female companion Capt. C— and myself upon the banks of the Allegheny river, just as we were viewing some laurels that garnish an impending rock hanging over the river, two Indian men and an Indian woman came suddenly upon us. (48)

In keeping with literary portrayals of surprised women at the time, and for years to come, Caroline screams and faints into Arl—ton's arms. Almost comically, the Indians approach and ask "if they could be of any assistance to Caroline," before adding to a likewise shaken Arl—ton, "Brother, if we have been your worst foes in war, we will be your best friends in peace," and they depart saying that "they were going to Pittsburg for the purpose of burying the hatchet" (49).

Notable about this second major reference to the Native American threat is the way in which Arl—ton uses the fear of Indians as an opportunity to bring Caroline back into the patriarchal fold. Caroline's nature walks, which Arl—ton dismissively calls her "little excursions," liberate her both mentally and physically in a realm of beauty no European American man can master. Yet as soon as the mere possibility of Indian violence enters her mind, she faints on cue and literally falls into Arl—ton's protecting arms. Whether or not Arl—ton consciously benefits from the specter of Indian hostility, his relationship with Caroline does indeed strengthen because of it. In fact, he first realizes how deeply he loves her as she awakes in his arms. As she recovers her composure and "apologized for having been so weak" (49). Arl—ton follows by telling her such a reaction was "natural," perhaps hinting that her outdoors-inspired independence and self-assuredness were unnatural for a woman. Building upon the idea of establishing just what is and what is not natural to Arl—ton, he also admits that as she awoke in his arms he was overcome with affection for her, writing "I felt in the returning pulsations of her hand, the high beating tones of *Nature* vibrate through every part of my frame; and in this ecstatic moment, when language was mute, and when Caroline's eyes spoke the most ineffable things, I was lost in the Elysium of intoxicated desire" (49, emphasis added). For Arl—ton, Caroline and his feelings for her are never more natural than when the empowerment of her nature walks is replaced by her need for his protection. Her Arcadia gives way to his "Elysium of ... desire." For as Imlay makes clear, in each case the characters experience a wilderness constructed and defined by their own imaginative and creative limits, and in no instance is that wilderness entirely accurate. Caroline's is a pastoral fantasyland; Arl—ton's is Caroline herself. As John Seelye argues,

> where Imlay may have thought to promote the Ohio Valley as a zone of social transformation, his pastoral descriptions tend to put forth a distinctly erotic image, recalling Thomas Morton's New-English Canaan and verifying Annette

> Kolodny's lay of the literary landscape in America as a terrain defined in feminine terms with seductive connotations. Thus when we first catch sight of Imlay's heroine, Caroline T—n, she is described as part of the scenery against which she poses ... Comparing a pretty woman to the delightful aspects of nature is a common literary touch, but through a series of repetitions, Caroline becomes a virtual personification of the "innocent realm"' toward which she is heading.[29]

That eventually Caroline's fear comes true when she is indeed kidnapped by hostile Indians and must rely upon the heroic Arl—ton to save her from whatever fate they had planned demonstrates how Imlay toys with gender expectations while also commenting upon the land as connected to racial violence. Caroline and company are told that the Indians are too distant to form a threat, and once they encounter actual Indians they are told that, having made peace with the whites, these particular Indians are no longer foes. Then, when finally Caroline is briefly captured by a party of hostile Indians, these much-feared enemies are, for the most part, easily handled, with Arl—ton quickly reclaiming Caroline for himself, if not for her actual freedom.

Caroline is, after all, never as free as her walks in the woods make her feel; she is either under the control of the European American patriarchy or under the control of the Indians. This nod toward women's rights melds well with Imlay's point when he reminds his readers that Indian captivity was not always viewed as a horrible fate for some women: "I have been told of instances, where women have been treated with such tenderness and attention by them [their Indian captors] that they have from gratitude become their wives" (204). Verhoeven and Gilroy write, "Imlay seems quite sympathetic to those women who acculturated and remained with their captors; indeed, as recent criticism has demonstrated, captives often found freedom from the rigid gender roles of white society, which deemed them helpless and sexually vulnerable."[30] While Imlay's mention that some kidnapped European American women welcomed assimilation does seem sympathetic, his sympathy seems less certain in terms of Caroline's captivity. In fact, rather than function as an avenue for liberation from the patriarchy, her captivity serves instead to put her back in her place as a woman fit to marry Captain Arl—ton. Additionally, Caroline's experience reveals the degree to which the environment serves as a buffer zone between the two systems of control, between the European American social structure and the perhaps terrible future she would experience in the hands of the Native Americans. As with most early American texts, the land's indigenous population is intimately linked with the environment in that they essentially become signifiers for each other. Both are assimilated, degraded, and reshaped to fit the purposes of the colonial American agenda, and perhaps nowhere in *The Emigrants* is this situation made more apparent than in Caroline's captivity.

---

[29] Seelye, "The Jacobin Mode in Early American Fiction: Gilbert Imlay's *The Emigrants*," 208–9.

[30] Verhoeven and Gilroy, "Explanatory Notes," *The Emigrants*, 292–3 n. 2.

The events leading up to her kidnapping and then to her rescue by Arl—ton are worth examining in relation to how the land itself factors into which party—the Indians or the whites—control this young woman and her future. The T—n family arrives by boat at Louisville, stopping where the falls of the Ohio River force them to disembark. Caroline's spirits suffer from a combination of homesickness for recently departed friends, lovesickness for Arl—ton, and an actual physical sickness that has kept her bedridden for a short time. While none of these ailments place her in significant danger, they do combine to create in her a mild depression. Once well enough to venture outside, she attempts to assuage her melancholy in the way she knows best, with a nature walk. The drama begins when Caroline selects a somewhat perplexing location for this outing, of which her uncle, in a letter breaking the news of her capture, writes:

> to amuse her thoughts she crossed this morning into the Indian country, unattended, except by the barge-men, who were to return for her at 3 P.M. and her maid, to view the falls from Silver Creek, where she was when the maid, who was returning back from Clarkeville where she had been for her mistress's glass, that had been left by accident, saw the savages carry her off. (192–3).

Imlay sets a pastoral crime scene for Caroline's abduction, and this scene, along with a key piece of physical evidence, illuminates his point about the difficulty women faced in obtaining a modicum of liberty and happiness in a world in which they are—like the land itself—owned either by the whites or by the indigenous population. In choosing to go for a stroll in what she knew was dangerous territory, Caroline attempts to assert her own will on the situation. Given her well-known predilection for the most Arcadian and sublime of views, it is unsurprising that she selects the falls as her destination and takes along "her ... glass," which in the course of her abduction falls to the ground for her maid to find. According to Verhoeven and Gilroy, this "glass" is "probably a Claude glass, a slightly convex blackened mirror, popular in the eighteenth century as a device to view landscapes."[31] Taking its name from the celebrated French landscape painter Claude Gelée (1600–1682), this mirror would apparently afford its owner the power to render actual natural scenes in a manner that approximated the style of Claude himself. Holding the glass toward her face, the owner of a Claude glass would turn her back on the view in question and look into the darkened reflection found in the glass. As Verhoeven and Gilroy contend, "the Picturesque tourist would see a 'composed' image in finely graduated, harmonizing tone, theoretically reminiscent of the landscape paintings of Claude."[32] Imlay, by stressing that Caroline was holding this device in the moments before she was kidnapped, demonstrates the danger women face when they turn their back upon reality in preference to a romanticized view. In Caroline's case she appears to have been so absorbed in the artificially pastoral-tinted scene that she was blind to the approach of the hostile Indians. In the pages that follow her abduction Caroline becomes firmly and forever returned to the

---

31 Verhoeven and Gilroy, "Explanatory Notes," *The Emigrants*, 290 n. 3.

32 Verhoeven and Gilroy, "Explanatory Notes," *The Emigrants*, 290–1 n. 3.

European American control of her brave and love-struck Captain Arl—ton, who expertly manages to make away with her without killing any of the Indians. Imlay makes sure to note that the young woman was not raped or otherwise molested by the Indians, having their friend Mr. Il—ray relate to Caroline's uncle that during her captivity "she never felt in the least harassed, or alarmed for her safety" and "that the Indians treated her the whole time with the most distant respect, and scrupulous delicacy" (203). This information not only assured the reading public that the Indians did not quite live up to their savage reputation, but also indicates that Caroline remains an unspoiled prize for Arl—ton.

In fact, Arl—ton's prize is not only the hand of the lovely Caroline but also a considerable portion of the land as well. In her final letter of the novel, Caroline—now addressing herself as Mrs. Arl—ton—informs her sister of her marriage and of their new home on the very land that served as the setting for both her most picturesque outings as well as her brief capture. In one of the book's most hyperbolically pastoral and anthropomorphic passages, Caroline explains at length that,

> Captain Arl—ton has purchased a tract of country, sixteen miles square, adjoining to the rapids, which form a stupendous cataract.
>
> This tract is bounded by the Ohio on the west; and here the expansive river displays in varied pride the transparent sheet, that gushing, shoots impetuous over its rocky bed; which, as if in a rebellious hour had risen to oppose its genial current, presents a huge, but divided barrier; and while nature seems to scorn its feeble power, the repercussive thunder proclaims her triumph, and the ethereal hills on the adjacent shore give lustre to the rising moisture. (245)

The newly minted Mrs. Arl—ton has clearly replaced her figurative Claude glasses, and, now that she is married to the new owner of the land, her pastoral fantasy may continue undisturbed, even if the land itself will not. For as Caroline continues with her description of Arl—ton's considerable landholdings she also provides a glimpse of how the land will fare under his stewardship:

> The country on the opposite shore, over hung with woods, is not less rich in variety; but as it remains yet uninhabited, we have the charms of cultivation contrasted by the beauties of wildness.
>
> This body of land, Arl—ton has parceled out into a number of lots, which are in part settled, and the remainder are settling, he having reserved six for himself and those friends who may in future wish to join us. (246)

The book thus ends with Caroline and Arl—ton able to have the best of both worlds. Following in the tradition of so many real life land speculators, he will divide, lease, sell, cultivate, and otherwise settle the land, profiting from its transformation into a garden, one made more pleasing by the wilderness that—at least for the moment—still drapes the banks on the opposite side of the river.

Edgar Huntly's Subterranean Ecogothic

While neither Imlay's *The Emigrants* nor Bleecker's *The History of Maria Kittle* have, at least to this point, attracted a large amount of critical study, Charles Brockden Brown's novel *Edgar Huntly*, due to a renewed interest in Brown's work in general, has benefited from significantly more scholarly attention.[33] Yet the body of critical work devoted to *Edgar Huntly* has predominantly concerned matters unrelated to the book's portrayal of the late eighteenth-century Pennsylvania wilderness itself. The scholarship that does touch upon Brown's portrait of what is a decidedly psychic and disturbingly imagined natural world, quite understandably does so in association with his literary treatment of the Delaware Indians, who form a key component of the novel's plot. The story is, after all, in large part about the problem of European American anxiety over the dispossession of the Delaware Indians and the colonization of the land that was their home. This theme becomes entwined in the larger story of Edgar Huntly's search for the killer of his best friend, Waldegrave, while also attempting to help a grief-stricken Irish immigrant named Clithero Edny (a man who may or may not have something to do with that murder). All of this plays out while Huntly suffers from his own personal demons, which in part manifest as a dangerous tendency to sleepwalk. This summary, however, does not really encapsulate the novel, in part because in true gothic fashion, Brown spins a chaotic narrative web of mysterious figures, uncanny doubles, tangential tales, hidden manuscripts, secret boxes, grotesque figures, and bloody violence. In much the same way as the Panther Captivity before it, the novel melds gothic tropes with the new American threat of the monstrous Indian, who in ghostlike fashion inhabits a wilderness full of strange threats and sublime beauty. It is a thrilling, confusing, disturbing, and frustrating tale, and it has the distinction of serving as the one of the major launching points for the American frontier adventure, representing a confluence of Indian captivity narrative, gothic novel, pastoral fantasy, and veiled social commentary.

Few if any pre-1800 American fiction writers were as aware that they were helping create a new national literature as Brockden Brown. Brown's cognizance of the problems inherent in attempting to fashion an American fiction from salvaged components of European influence and the day-to-day of American existence is easy to recognize in his oft-quoted prefatory remarks to *Edgar Huntly*, in which he famously writes, "America has opened new views to the naturalist and

[33] I make no attempt to list all of the recent Charles Brockden Brown scholarship, but it includes a growing author's society, which may be found at http://www.brockdenbrownsociety.ucf.edu, and a new multivolume edition of the *Collected Writings of Charles Brockden Brown*. The role of the environment in *Edgar Huntly* has been of much interest to scholars for a long time, with some of the best studies arising in the 1980s. See Berthold's "Charles Brockden Brown, *Edgar Huntly*, and the Origins of the American Picturesque" and "Desacralizing the American Gothic: An Iconographic Approach to *Edgar Huntly*," 127–38; and Lueck's "Charles Brockden Brown's *Edgar Huntly*: The Picturesque Traveler as Sleepwalker," 25–42, to name just a few.

politician, but has seldome furnished themes to the moral painter."[34] Brown wears his intentions openly as he goes on to state that "It is the purpose of this work to profit by some of these sources; to exhibit a series of adventures, growing out of the condition of our country" (3). For Brown a new American literature must incorporate the natural condition of the country and that condition was incomplete without the inclusion of what he calls, "The incidents of Indian hostility, and the perils of the Western Wilderness"; moreover, he warns that "for a native of America to overlook these, would admit of no apology" (3).

The wilderness of Brown's *Edgar Huntly* owes much of its dark, romantic portrayal to the fictionalized or exaggerated environments found in some of the more sensational examples of American Indian captivity narratives. For example, the similarities between the landscape imagery of Brown's 1799 novel and that of the 1787 Panther Captivity indicate that Brown had probably read and become influenced by such texts. He makes it clear almost from the start that the pastoral natural world of texts like Imlay's *The Emigrants* has been replaced by an almost uniformly threatening landscape. Brown simultaneously mirrors the chaotic mindset of the title character while emphasizing the American Gothic wilderness as a dark and labyrinthine realm.

The book opens with young Edgar Huntly's nighttime ramble through the wilds of frontier Pennsylvania. Grieving for his recently murdered friend Waldegrave, Huntly finds himself once again drawn to the scene of his friend's death. He writes that upon walking home that night

> The impulse was gradually awakened that bade me once more to seek the Elm; once more to explore the ground; to scrutinize its trunk. What could I expect to find? Had it not been an hundred times examined? Had I not extended my search to the neighbouring groves and precipices? Had I not pored upon the brooks, and pryed into the pits and hollows, that were adjacent to the scene of blood? (8)

Brown thus offers a wilderness murder mystery. Huntly serves as an amateur sleuth who must traverse the imposing and treacherous forest to find clues to his friend's death. He also displays hints that he himself is not entirely of sound mind. His thinking is often as confused and obscure as the wilderness he navigates to the crime scene:

> I was familiar with the way, though trackless and intricate, and I climbed the steeps, crept through the brambles, leapt the rivulets and fences with undeviating aim, till at length I reached the craggy and obscure path ... In a short time, I descried through the dusk the widespread branches of the Elm. This tree, however faintly seen, cannot be mistaken for another. The remarkable bulk and shape of its trunk, its position in the midst of the way, its branches spreading into an ample circumstance, made it conspicuous from afar. My pulse throbbed as I approached it. (9–10)

---

34 Brown, *Edgar Huntly*, 3. Subsequent references appear parenthetically in the text.

As this description indicates, Huntly's night hike to the murder scene under the elm tree takes place in a wilderness not meant to resemble the actual Pennsylvania forest. It is rather the emblem of the muddled, grief-stricken, and ultimately vengeful thinking of the young Huntly. At the beginning of this tale Huntly stands in the middle of the road, approaching a monolithic natural signpost, the majestic elm, an imposing remnant of the old growth forest. Yet instead of serving as an inspiring site of natural grandeur, the tree functions as the marker of a crime scene, even as a sort of natural grave marker for the murdered Waldegrave (whose name means "grave in the woods").

That this ominous elm calls forth associations with Penn's Elm is no coincidence. Syndey Krause persuasively argues that this elm tree

> is the crucial landmark from which action springs and nexus of the collision and divergence of major plot lines (murder, guilt, pursuit, sleepwalking, and Indian warfare)—this great Elm functions as a psychological magnet on Huntly and Edny, fateful as much as "fatal." The tree not only becomes a motival icon for Brown, but it evokes a cumulative significance beyond itself by historical linkage with a comparably central image in the founding of Pennsylvania and, specifically, Philadelphia—heart of the nation to be. Namely, it evokes the long honored symbolism of the legendary "Treaty Elm," under which, tradition has it, William Penn ... negotiated a treaty of friendship ... with the Lenni Lenape.[35]

So just as the former arboreal landmark was meant to represent reconciliation and peace between the Native Americans and Penn's Quaker settlers, the "haunted Elm" (14) of Brown's novel works as its dark inverse. It is the location where a disgruntled Lenni Lenape Indian arbitrarily murdered Waldegrave out of hatred for the oppressive encroachment of the European American settlers. Thus in this intriguingly chaotic novel a master trope of the American Gothic emerges in one of its first fully realized incarnations: the fear of the return of the oppressed racial other through acts of violence in a terrifying wilderness. This trope emerges as one of the main thematic engines of the book, and in this case it is fulfilled by both the Lenni Lenape and by Huntly himself, depending on whether he is ultimately justified in his actions (he is after all, in this epistolary novel our only source of information, and he, on occasion, proves himself to be less than reliable). In fact, Brown—taking cues from the conventions of European Gothic—employs Huntly as the means for exploring a host of sociopolitical and nationalistic questions plaguing the early republic psyche: At what point must we curtail liberty for the greater good? What right do European Americans have to the land upon which they founded their new nation? What is the place of the racial other in the composition of the new country, and how can we navigate the hypocrisy inherent in a nation that professes to be the home of the free when so many are literally (as well as metaphorically) enslaved? In a democracy, how can we trust to the sanity and integrity of our leaders and of those who place them in power? Brown employs the

35 Krause, "Penn's Elm and *Edgar Huntly*: Dark 'Instruction to the Heart,'" 464.

labyrinthine wilderness of his book to pose, if not actually answer, such questions, and it would seem that at the core of each is the human relationship with the land, with that purloined resource upon which virtually all hopes of prosperity reside.

Yet Brown also takes this metaphor a step further by having the environment mirror the confused mind of Huntly, and in its dark perplexing narrative reflect the bewildered relationship that both the title character and the settlers of the nation suffered when attempting to solve this puzzle that forms the futures of the land, the European Americans, and the tragically wronged Lenni Lenape. As Dennis Berthold explains, Brown

> recognized that the descriptive conventions and emotional associations of picturesque travel provided visual analogues to the romantic hero's psychological state; in this way, picturesque aesthetics mediated between nature and the ego, the "eye" of vision and the "I" of selfhood, and so helped balance the conflicting claims of wildness and disorder, on the one hand, with those of civilization and order on the other.[36]

Berthold's take on Brown's adaptation of the picturesque certainly works, but it necessarily deals mainly in surfaces rather than depths, with an exposed land and an exposed mind, a tendency that falls short in examining the subterranean possibilities—physical and psychological—found in *Edgar Huntly*. After all the Gothic is about that which is buried and in the dark, in which case the "eye" is rendered either useless or unreliable, and the "I" of the ego (as we see in Edgar's transformation in the dark pit) becomes muddled as well.[37]

In the case of *Edgar Huntly* the cultural guilt born of the injustices against the environment and the Native Americans (so often conflated with the land) are buried both in the landscape and in Huntly's much-confused mind. These thematic undercurrents find their way into Brown's description of the geography and geology of the Pennsylvania wilderness. In fact, Brown displays a penchant for creating highly charged instances of subterraneous passageways; vertigo-inducing forest cliff sides; and lurking, monstrous panthers (as well as Indians who take on

---

[36] Berthold, "Charles Brockden Brown, *Edgar Huntly*, and the Origins of the American Picturesque," 63. Like Berthold, I argue that the landscape mirrors Huntly's mental state, and it would seem we are in good company. In his book, *Story Line: Exploring the Literature of the Appalachian Trail*, Ian Marshall points out, "almost every commentator on *Edgar Huntly* draws some connection between the landscape and Edgar's state of mind," a critical tendency Sydney J. Krause traces back to 1819, in a review by E. T. Channing, 261 n. 8; also see Krause, "Historical Essay," 392.

[37] Ian Marshall comes closer to this realization when he writes, in *Story Line: Exploring the Literature of the Appalachian Trail*,

> Because they find something that appeals to their psyches in the gloomy woods, Huntly and Edny spend a lot of time there, and so they know the woods well enough to find their way around, even in the dark. Even, evidently, in their sleep. When the landscape becomes disorienting, it both reflects and causes Edgar's mental confusion. (136)

animalistic traits); all of which appear in the novel as connected to the subsurface relics of colonial conflict and environmental abuse. One of the best examples of how this enters the narrative is when Huntly stalks the mysterious Clithero (whom he erroneously believes has killed Waldegrave) through the nighttime wilderness only to see him disappear into a cavern entrance. The next day Huntly plans to again pursue the strangely predictable nighttime ramblings of Clithero (who it seems is sleepwalking, which Huntly takes as evidence of his guilt in the murder of Waldegrave). Huntly, sorting through his plans, writes, "But what proof had I that the same rout would be taken, and that he would again inter himself alive in the same spot?" (22). Brown invokes what would, in the hands of that later master Edgar Allan Poe, become the common Gothic trope of the fear of live burial. As it turns out, such a fear is easily realized in Brown's dark forest because, as Huntly writes,

> The basis of all this region is *limestone*; a substance that eminently abounds in rifts and cavities. These, by the gradual decay of their cementing parts, frequently make their appearance in spots where they might have been least expected. My attention has often been excited by the hollow sound which was produced by my casual footsteps, and which shewed me that I trod upon the roof of caverns. (22)

Like the often suspect reasoning of the narrating Huntly, the landscape upon which this story plays out is also suspect and unsound, veined with dark corridors of decay, and creating as a result an underground labyrinth simultaneously romantic and disturbing. The land that serves as the stage for Huntly's odd narrative is literally porous, rotten—a subterranean limestone sublime. Yet even above ground the wilderness is maze-like in its deeper recesses, and in following Clithero from the outskirts of the forest into an even deeper wilderness beyond, he writes,

> The track into which he now led me was different from the former one. It was a maze, oblique, circuitous, upward and downward, in a degree which only could take place in a region so remarkably irregular in surface, so abounding with hillocks and steeps, and pits and brooks as *Solebury*. It seemed to be the sole end of this labours to bewilder or fatigue his pursuer, to pierce into the deepest thickets, to plunge into the darkest cavities, to ascend the most difficult heights, and approach the slippery and tremulous verge of the dizziest precipices. (23)

A key word in this vertiginous passage is *bewilder*. Clithero is intent (or at least it seems so to Huntly) upon bewildering the pursuing Huntly by taking him into a forest so fraught with topographical chaos that it becomes essentially innavigable and impenetrable even to precise description. Similarly, Brown, throughout the novel, takes the reader on a comparable Gothic journey, inviting us to chase Huntly from one dizzying passage (dizzying both as written and otherwise) to the next as the terrors of the natural world intertwine with the anxieties of colonial guilt. The result is an accretion of literary *bewilderment* that makes us as unsure of our own reading of the novel's twists and false turns as Huntly ultimately transforms into a figure of misguided vengeance.

This environmentally induced transformation occurs in large part during one of the critical set pieces of the novel, the passage in which a formerly sleepwalking Huntly awakes to find himself trapped in a pit inside a cavern. Much has been written about this scene, but few if any critics have connected it with Brown's larger project of portraying Huntly as a man plagued by a form of extreme bewilderment brought about in large part by his frustrated attempts at plumbing the wilderness to find meaning in Waldegrave's death and in his own life. Huntly, who seems to transfer his own social, familial, and economic anxieties into a concern for the safety of the enigmatic Clithero, apparently takes on so much worry and stress that he, like Clithero, begins to sleep walk, a condition that puts him in terrible peril. The moment in question begins in chapter 16 when Huntly writes that what follows is a "scene ... pregnant with astonishment and horror" (158). It is a scene that when viewed within Brown's larger environmental project represents (for all of its subterranean terrors) the pinnacle of Huntly's bewilderment, the consummation of Brown's American environmental Gothic. The famous scene begins with Huntly, whose last memory is of falling asleep in his bed at home, awakening to severe body pains in a cold realm of total darkness. At first he thinks he has gone blind, but he soon realizes he is actually in a place almost entirely devoid of light. "I was supine upon a rugged surface and immersed in a palpable obscurity" (161). He feels around and finds, of all things, a tomahawk on the ground, and wonders, "What dungeon or den had received me, and by whose command was I transported hither?" (161). He touches a wall and, following its course, finds he is "merely going round the walls of a vast and irregular apartment" (161). Though he is conscious, the sensory deprivation he experiences forces him to question whether he is really awake ("I existed as it were in a wakeful dream" [161]) or if he has not fallen victim to the Gothic trope of live burial: "Sometimes I imagined myself buried alive. Methought I had fallen into seeming death and my friends had consigned me to the tomb, from which a resurrection was impossible" (162). Huntly, who soon realizes he is trapped in a pit within a dark cavern, attempts to climb out but cannot. Madness begins to take over. He becomes voraciously hungry and thirsty as he begins to figuratively regress into an animalistic state. He eats part of his shirt to assuage his hunger pains, and finally resolves to bury the tomahawk within his own chest to end his misery. Just as he is about to take his own life, he notices two glowing eyes in the dark pit and realizes that they are the eyes of a panther about to spring upon him. Huntly throws his newfound tomahawk and kills the panther. He then satiates his intense hunger by drinking the blood of the dead animal. Eventually Huntly makes his way out of the pit and then blindly through the dark cavern until a dim light leads him to an exit.

The role of the cavern in *Edgar Huntly* is radically different from that of the cave in the Panther Captivity. While both of these subterranean realms represent a sort of inherent Gothic darkness built into the natural landscape, the cave of the Panther Captivity ultimately becomes the woman's home, and as such something of a symbol of female agency in the otherwise treacherous wilderness. *Edgar Huntly*'s cavern, on the other hand, remains a labyrinth of fear, stained by eternal night. Rather than empower those who enter it, this cavern attracts the sleepwalking Huntly, nearly killing him, only to release him from one darkness into another. Brown's dungeon-like cavern functions as a perversion of Plato's "Allegory of the Cave": those who emerge from it do not enter a world of enlightenment but rather one of further confusion and violence. In fact, young Huntly drags himself from the treacherous cavern not into daylight at all. Rather, he finds it is nighttime and the light that seemed a beacon pointing his way to salvation is instead the campfire of several Lenni Lenape men, who appear to have taken a young woman as their prisoner. The narrative shifts dramatically the moment Huntly exits the cave, and with this shift comes a change in the role and portrayal of the natural world, which becomes inextricably conflated with that of the Indians he confronts upon his exit. Huntly frees the captive woman, but decides to also kill the Indians. As with so many of the preceding environmental texts detailed in this chapter, *Edgar Huntly*'s portrayals of the natural world are intimately linked to the plight of the American Indian. It is, after all, the Lenni Lenape who have suffered decades of abuse from the assimilationist policies of the whites. As is the case with earlier environment texts, the land figures as the site of Indian captivity and racial violence. Huntly manages to free the captive woman without bloodshed, but unlike the clear-headed Arl—ton of Imlay's *The Emigrants*, he does not race back to the settlements for safety. Instead, he insists upon murdering the Indians despite his uncertainty of their actual intentions (though given the context of Lenni Lenape–settler relations, he is not necessarily off base in assuming the girl has been kidnapped).

The fact that he resorts to such violence only on supposition speaks to the way Brown deftly interweaves a prescient racial message into the action of young Huntly, who seems largely immune to self-doubt when it comes to dispensing violence against either the Indians or those species of wild animals he deems as pests. Indeed, the wild animal that Huntley seems to most malign is the panther, a creature that in his description appears to reflect the racist stereotypical description often levied against Native Americans. Huntly relates that along with rattlesnakes, he abhors the panther, which he describes as follows: "His grey coat, extended claws, fiery eyes, and a cry which he at that moment uttered, and which, by its resemblance to the human voice, is peculiarly terrific, denoted him to be the most ferocious and untamable of that detested race" (124). Elsewhere he describes the panthers as "sanguinary spoilers" (124), as "enemies of man" (124), and as "savage" (126). As his various descriptions of the panthers indicate, Huntly thoroughly anthropomorphizes these creatures while aligning them with the Native Americans. In Brown's American environmental gothic, these two sets of American monsters share more than the traits imposed upon them by Huntly, they both—through European American environmental destruction and concerted eradication efforts—have become an increasingly scarce, yet spectrally present, figure in the wilderness. Brown links them to the natural world while also commenting upon how Europeans Americans simultaneously demonize and destroy any entity that they cannot fully control. In Huntly's mind the panther and the Native American become interchangeable threats to European American control of the land and must therefore suffer elimination.[38]

Brown's portrayal of the environment in *Edgar Huntly* stands as one of the most significant, if still prototypical, works of early American environmental fiction before Cooper's *The Pioneers*. The book itself is positioned in a key literary space between the increasingly fictive and popular captivity narratives of the late eighteenth and early nineteenth centuries and the novel as it would become as the nineteenth century wore on. Brown's portrayal of the environment as a bewildering, at times subterranean, world of nightmares, makes it perhaps the first of what would become a host of works that are simultaneously environmental and Gothic. Brown's novel is, after all, unapologetically modeled in the Gothic vein, and thus it displays what were by the end of the eighteenth century some common Gothic tropes, most of which may be found in environmental gothic texts up through the present day (for example, William Faulkner's *Go Down Moses*, James Dickey's *Deliverance*, and Toni Morrison's *A Mercy*).

As a gothic text, Brown's novel by default embraces romantic ideas about the natural world, but it does so by replacing a shopworn pastoral environment with a dark and treacherous wilderness, where mankind, for all its power, is never quite in control. Finally, the book's treatment of young Edgar Huntly's denigrating conflation of American Indians and animals (especially panthers) makes for a disturbing subtext in which Brown forwards the idea that the colonization of the American wilderness has left behind it a wake of oppression that just might come back to haunt the American cultural psyche. If, as Eric Savoy contends, the key to American Gothic is "the return of the repressed Other," then Brockden Brown forwards—perhaps for the first time—what would become a common theme of the American environmental novel: the idea that European American colonialism not only woefully oppressed the Native Americans, but also did the same to the land.[39] The portrayal of this shared suffering creates a warped cultural intimacy between the indigenous population and the environment, resulting in what has become a common trope of a spectral people roaming a haunted wilderness.

Conclusion

It is enticing to think of these earliest works of environmental fiction as emerging on the formative American literary scene in a meaningful progression, moving from, say, the most basic of works to ever more sophisticated attempts at formulating an environmental imagination, but that perception would be mostly incorrect. While the genre of American fiction does indeed mature between 1787's Panther Captivity and 1799's *Edgar Huntly*, their incorporation of the natural world is not

38 For a more developed examination of the connection between environmental concerns and panthers in antebellum American fiction see my "'The Base, Cursed Thing': Panther Attacks and Ecotones in Antebellum American Fiction."

39 Savoy, "The Face of the Tenant," 14.

so easily tracked. After all, even the earliest of these texts presents an amalgam of environmental sophistication and well-established pastoral conventions.

The 1780s and 1790s thus represent key decades in this advancement, with some of the first steps toward the creation of an imagined natural world appearing in the Panther Captivity, that odd narrative made psychological puzzle. By the 1790s, books as distinct in subgenre as they were in origin began to incorporate glimmers of an environment that cast light beyond mere Romantic posturings. We see this emergence in the sentimental fantasy of Imlay's *Emigrants*, a book almost improbably set in Kentucky's Ohio River Valley. Likewise, Bleecker's *History of Maria Kittle* and Charles Brockden Brown's *Edgar Huntly* each present a melding of genres and modes, drawing together the tropes of the sentimental romance and the Gothic—planting each in the American literary soil.

Roderick Nash writes that "for the most of their history, Americans regarded wilderness as a moral and physical wasteland fit only for conquest and fructification in the name of progress, civilization, and Christianity."[40] While this trend certainly played out in terms of real-world environmental practice, by the early nineteenth century, America's literature, especially its fiction, began to shift toward a more environmentally appreciative (if still Romantic) mode. It would seem that early nineteenth-century fiction allowed readers to appreciate the environment by proxy, in the guise of the environmentalists they wished they could be. As the next chapter illustrates, these concerns were increasingly found in the juvenile literature of the period as well, and by the 1820s and 1830s, with the success of books like the Peter Parley series, we witness a proto-conservationist trend that would seed those same young minds who would grow up to participate in the cultural flowering of the American Renaissance.

---

[40] Nash, *Wilderness and the American Mind*, vii.

Chapter 3
Juvenile Environmental Literature

Do you know what great forests there are in America? They were much greater formerly, than they are now.

—Samuel Griswold Goodrich (aka Peter Parley), *The Child's Botany* (1828)[1]

In the previous chapter we saw how U.S. environmental fiction developed in part from the already popular captivity narrative tradition, but as this chapter will demonstrate, adult fiction was only part of the early republic's book market. Running concurrently with the adult book trade was a strong transatlantic demand for juvenile literature. Considering its pronounced influence on the nation's long-term intellectual development, juvenile literature was undoubtedly one of the strongest factors in the development of a distinct American environmental consciousness.

Writing in 1933, the legendary book collector A.S.W. Rosenbach argued that children's books "more than any class of literature ... reflect the minds of the generation that produced them. Hence no better guide to the history and development of any country can be found than its juvenile literature."[2] Rosenbach's claim remains a compelling reminder that our culture rarely wears its values, hopes, and fears so openly as in the literature intended for the instruction of its children. With this idea in mind, it is productive to examine the extent to which environmental concerns, anxieties, and blindnesses emerge in the early republic's juvenile literary culture, a culture that was, in the years following the American Revolution, very much in its own juvenile stage. Americans of the early republic were increasingly avid readers, and what they read had a profound impact upon their children's ways of thinking about the natural world.[3]

When adults wrote, printed, sold, bought, read, reviewed, and shared these books, they were teaching the next generation to either accept or refute the land in which they had been born. And that land was changing fast. As popular as literacy became in the fledgling republic, there was—aside from direct exposure to the natural world itself—nothing more influential to the ways children thought

---

1 Goodrich, *The Child's Botany*, 14.

2 Rosenbach, *Early American Children's Books*, xxvi–xxvii.

3 There are relatively few studies of early American juvenile literature, and even fewer examine these texts from an environmental perspective. Ecocritical studies of children's literature from later periods are more common, and one particularly helpful anthology is *Wild Things: Children's Culture and Ecocriticism*, edited by Sidney I. Dobrin and Kenneth B. Kidd; see especially Maude Hines's article, "'He Made *Us* Very Much Like the Flowers:' Human/Nature in Nineteenth-Century Anglo-American Children's Literature," which offers an instructive look at nineteenth-century botanical children's literature.

about a country that was on the brink of an industrial expansion than the books that were themselves evidence of that expansion. In taking a closer look at the literature produced at this time and read by millions of young people, we gain a better appreciation of the cultural climate that preceded the more pronounced development of American nature writing and environmental thought in the second half of the nineteenth century. If we ignore works that were most likely read by some of our greatest early environmental thinkers during their younger days, we risk missing an important factor in the development of the nation's perception of the natural world. After all it is likely that, once upon a time, young Cooper, Emerson, and Thoreau thrilled at reading some of these very books.

Origins

Sidney I. Dobrin and Kenneth B. Kidd argue that the ideology behind the literary portrayal of the environment in twenty-first-century children's literature

> is, at the most general level, twofold. On the one hand, there is the belief that children are innocent and/or virtuous, in keeping with the romantic philosophy of Rousseau and other advocates of "original innocence." ... On the other hand, the child is still assumed to be devoid of content ... The child thus has no necessary connection with nature, no experience or understanding of it."[4]

In the pre-Romantic period, which in colonial North America was marked by the cultural dominance of Puritan thinking, children were not born innocent; they were vessels of original sin just like their parents. And as far as what subjects should fill children's minds, it was clear that the only matters worthy of such trouble were of religious instruction. Unsurprisingly the environmental sophistication of juvenile literature mirrored that of its parent culture in general, and as we will see, with the dawn of Enlightenment thinking came a profound shift in the availability and environmentality of American juvenile literature.

As with much of early American culture, the first American juvenile narratives were not really American at all—they were British. And in the beginning they were—with the exception of the occasional mention of a wild animal—unconcerned with issues related to the natural world. In his bibliography of American children's books published before 1821, d'Alté Welch notes that the first children's book published in the colonies was a reprint of the British author James Janeway's *A Token For Children: Being An Exact Account of the Conversion, Holy and Exemplary Lives and Joyful Deaths of Several Young Children*, which was published in Boston in 1700 and to which Cotton Mather appended a section entitled, *A Token for the Children of New England*. Welch characterizes Janeway's book as a, "popular compilation of stories of over-pious, preaching children, who died at an early age

---

[4] Dobrin and Kidd, "Introduction," in *Wild Things: Children's Culture and Ecocriticism*, 5–6.

of some unpleasant disease."[5] The people of colonial Massachusetts, whether because it spoke to the Puritan mindset or because it was one of a very limited set of options, enjoyed the book, and it went through at least 29 American editions by 1816. The second narrative meant for juveniles that landed on the bookshelves of colonial America was Thomas White's 1702 *A Little Book For Little Children*, which Welch describes as "replete with preachments and includ[ing] harrowing accounts of tortures taken from Foxe's *Book of Martyrs*."[6] Despite their often-morose subject matter, works of juvenile literature were extremely popular, so much so that in many cases books once published in editions of several thousands are now known today only by one surviving copy. Some books were so popular that they were literally read to pieces by their adoring owners, with some editions entirely lost and others known today only by one or two worn copies. As Welch writes, early American children's books are "like fossils," that are "found today only because of a combination of factors which were favorable to their survival."[7]

As one might expect from the intensely pious Puritan culture, early examples of juvenile literature were anything but environmental. They were far more concerned with matters of the soul than with more earthly concerns, with teaching the children of New England to fear God rather than to love the land. Indeed, the Puritan conception of the natural world was largely antagonistic. In *Of Plymouth Plantation*, William Bradford famously paints the American environment as "a hideous and desolate wilderness, full of wild beasts and wild men."[8] The Puritan's environmental hostility coupled with a distrust of or outright disdain for anything resembling entertaining literature resulted in an atmosphere toxic to the creation of a proto-environmental narrative, especially outright fiction. While cohesive environmental stories were absent, by the early eighteenth century, environmental elements did, however, begin to make their way into publications for children. This inclusion of superficial environmental subjects makes sense because the main source of material for these books was, after all, the Bible, a book that is full of animals and other natural elements and that, in Genesis, presents readers with the concept of a lost environmental ideal.

An early example of how these biblical and occasionally secular environmental references made their way into books for colonial American children is the *New England Primer*. Published for decades to instruct children in matters of both spirituality and literacy, the *New England Primer* was the first American schoolbook. When placed alongside the Bible and Isaac Watts's popular hymnal for children, *The Divine and Moral Songs*, it completes the trinity of those books most likely to be read by American children until the late eighteenth century. With its dire religiosity the *New England Primer* was a hit with Puritan parents, who would read the books to their children until they were old enough to read them

5 Welch, *A Bibliography of American Children's Books Printed Prior to 1821*, xxi.

6 Welch, *A Bibliography of American Children's Books Printed Prior to 1821*, xxi.

7 Welch, *A Bibliography of American Children's Books Printed Prior to 1821*, xxii.

8 Bradford, *Of Plymouth Plantation: 1620–1647*, 62.

for themselves.[9] To provide some idea of how the book fared, Welch writes that in 1757 one printer is said to have churned out "one edition of 10,000 copies."[10]

While the *Primer* taught children the basics of reading and writing through a sustained regime of Christian doctrine, the natural world does seep in as an occasional learning tool. This tendency is especially apparent in its inclusion of a rhyming alphabet, which was well represented by a variety of animals. For example, a 1799 edition titled *The American Primer* contained, alongside pious entries such as that for A ("In ADAM's fall, / We sinned all"), more genial, animalistic examples, such as that for E ("An Eagle's flight, / Is out of sight").[11] In another section (among the pictures printed below the warning that "He who ne'er learns his A B C, forever will a blockhead be") can be found images of a range of creatures, wild and domestic, and sometimes even of a secular origin, whose names and images joined in the minds of children and set the model for the alphabet books kids read today.[12] The letter B corresponded with a bull; C and D were predictably portrayed by a cat and a dog; F was for fish, G for goat; and onward through Hog, Lion, Mouse, Nag, Owl, Peacock, Robin, Squirrel, Whale, and Young lamb. Z, however, was for "zany" (represented by a jester) in some editions and represented by the biblical figure "Zaccheus" in others.[13] It seems the zebra's reign would have to wait.

In the 1740s, British publishers such as Mary Cooper, and most famously John Newbury, began to print books that followed the new educational ideas of John Clarke and John Locke, who believed that childhood instruction should be fun and not an arduous task. Locke in particular believed that learning should be presented as an amusement and that the most educational of devices were those that seemed like toys. As Seth Lerer writes,

> The children's spaces of the Lockean universe are not filled with "tokens," or with signs of subsequent salvation. They are filled with objects, and Locke uses the word 'plaything' to connote not just the toys of the nursery or the bedroom, but the objects of experience that teach sensible and moral action.[14]

Responding to Locke's ideas, British printers began to create books that, as their titles suggest, were marketed as educational toys. This trend in making children's books more fun tended to make them more secular and allowed them to range into subjects outside of direct religious instruction.

---

9 Fessenden, *Culture and Redemption: Religion, the Secular, and American Literature*, 39.

10 Welch, *A Bibliography of American Children's Books Printed Prior to 1821*, xxii.

11 *The American Primer*, 14.

12 *The American Primer*, 11–12.

13 *The American Primer*, 11–12.

14 Lerer, *Children's Literature*, 106.

Nursery Rhymes, Natural Histories, and Animal Welfare Narratives

Some notable examples of the trend of marketing books as educational toys were two titles published by Mary Cooper. In 1743 she published *The Child's New Play-Thing: Being a Spelling-Book Intended to Make the Learning to Read a Diversion Instead of a Task*, which was reprinted by Edwards in Boston in 1750; and in about 1744 she published *Tommy Thumb's Pretty Song Book*, Volume II., which contained the first known printing of the "Mother Goose Nursery Rhymes." In 1780 T. Carnan, in taking over John Newbery's highly influential British children's book business, published these nursery rhymes as a stand-alone volume entitled, *Mother Goose's Melody; or Sonnets for the Cradle.*[15] Filled with catchy rhymes that relate the exploits of a talking sheep, a feline fiddle player, a moon-jumping cow, and many other nonsensical figures, this book—probably the most famous juvenile book ever—did not see print in America until well after the Revolutionary War, when, in 1785, the famed Worcester, Massachusetts printer, Isaiah Thomas, published the first American edition.

Mother Goose's Melody, along with another of Thomas's British reprints, *The Death and Burial of Cock Robin*, quickly became two of the most popular juvenile books in the early republic. *Cock Robin* alone went through more than 40 American editions between 1780 and 1809, and those editions are estimated to have been in the area of several thousand copies each.[16] As with the Mother Goose rhymes, *Cock Robin* is a rather nonsensical verse tale, opening with the famous lines, "Who kill'd Cock Robin? / I says the Sparrow, / With my bow and arrow."[17] As is true of children's books today, animals were exceedingly common figures in early American juvenile literature, and the creatures of the *Mother Goose* and *Cock Robin* books were hardly realistic. But after about 1788 a stream of books (many of them still British reprints) that took animals and the natural world a little more seriously began to trickle in from American print shops; these were the juvenile natural histories.

David S. Shields suggests that, "with its recondite dress and its fascinating images of prodigies and wonders, natural history radiated a mystery and novelty that stirred the popular imagination."[18] They were also, as Shields notes, exceedingly expensive to produce because they necessitated the inclusion of detailed illustrations of natural features, not to mention the amount of time and research that went into them. Indeed, natural histories were so notoriously

---

15 Welch, *A Bibliography of American Children's Books Printed Prior to 1821*, xxiv; Welch speculates that the late Newbery "probably had compiled or hired someone to compile the manuscript" prior to his death, and that in taking over Newbery's business, Carnan merely printed it; see Welch, *A Bibliography of American Children's Books Printed Prior to 1821*, xxx.

16 Welch, *A Bibliography of American Children's Books Printed Prior to 1821*, xxii.

17 *Cock Robin's Death and Funeral*, 14.

18 Shields, "The Learned World," 258.

expensive to produce that many never made a profit. Writing about the great cost of including illustrations in early American books, Georgia B. Barnhill writes that "In the early republic, popular interest in botany, geology, and zoology expanded greatly, but publications on these subjects faltered, owing to their large expense."[19] Barnhill explains that William Bartram's *Travels* (1791) set the model followed by other American works of natural history, including Benjamin Smith Barton's *The Elements of Botany*, William P.C. Barton's *Vegetable Materia Medica of the United States* (1818), Alexander Wilson's *American Ornithology* (1808–1814), and John Edward Holbrook's *The North American Herpetology* (1836–1840). "None of these works," writes Barnhill, "repaid their author-artists or publishers for their enormous investments of time and money. Each faltered over the expense of preparing hand-colored engravings or lithographed plates."[20] Nevertheless, it would seem that juvenile natural histories were profitable, which may have something to do with the fact that they at times reprinted sections and illustrations from the more expensive adult natural histories. For the most part, juvenile natural histories also tended to be shorter and adorned with less elegant cuts than their grown up counterparts. Some scholars estimate that juvenile natural history textbooks constituted only about 1 percent of the early republic's textbook market, but it would seem that these books also found a home in the non-textbook segment, a possibility supported by the fact that many are more akin to toy books than serious texts.[21] In all, juvenile natural histories ranged from lighter offerings that were essentially animal picture books to textbooks with detailed accounts of animals, plants, climate, geology, and astrology—sometimes all in the same volume.

One of the first juvenile books to offer young Americans a dose of natural history was *Jacky Dandy's Delight: or The History of Birds and Beasts*. This 1788 Isaiah Thomas reprint much resembles in format another popular juvenile natural history: *The Adventures of Tommy Trip*, which according to Welch, was "advertised in 1761 by John Newbery under the title *A Pretty Book of Pictures for Little Masters and Misses, Or Tommy Trip's History of Birds and Beasts*."[22] The first known American edition was published in 1796 when Samuel Hall of Boston reprinted it as *Pictures of Seventy-Two Beasts and Birds, with a Familiar Description of Each, In Verse and Prose, Published for the Benefit of the Lovers of Natural History, and for the Instruction of the Youth of America* (Figure 3.1).[23]

This book includes crude woodcuts coupled with brief verse and prose descriptions devoted to the baboon, fox, roebuck, rhinoceros, dog, and several other animals, both wild and domestic. The book views animals as objects of human exploitation, recreation, amusement, and derision. In describing the otter, it

19 Barnhill, "Transformations in Pictorial Printing," 439.

20 Barnhill, "Transformations in Pictorial Printing," 440.

21 Monaghan and Monaghan, "Schoolbooks," 306–7.

22 Welch, *A Bibliography of American Children's Books Printed Prior to 1821*, xxvii.

23 Welch, *A Bibliography of American Children's Books Printed Prior to 1821*, xxvii.

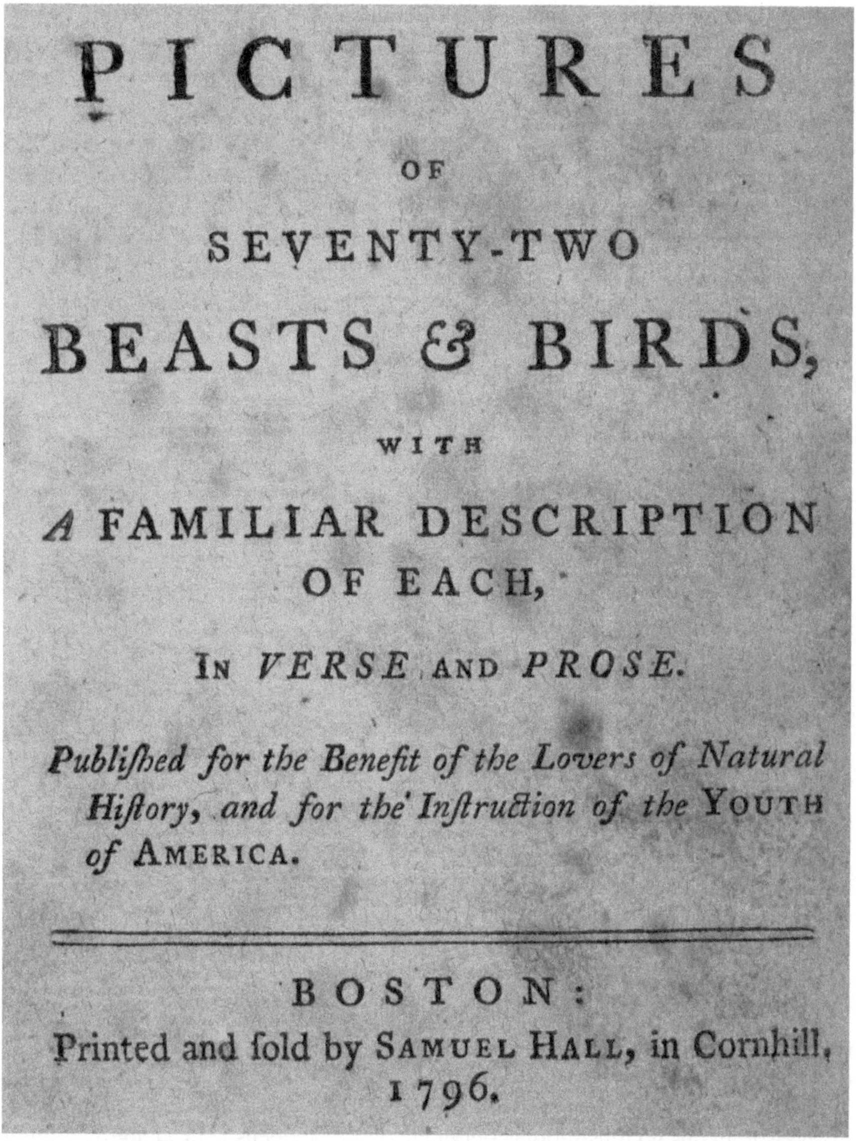

PICTURES

OF

SEVENTY-TWO

BEASTS & BIRDS,

WITH

A FAMILIAR DESCRIPTION
OF EACH,

IN *VERSE* AND *PROSE.*

Publiſhed for the Benefit of the Lovers of Natural Hiſtory, and for the Inſtruction of the YOUTH *of* AMERICA.

BOSTON:
Printed and ſold by SAMUEL HALL, in Cornhill,
1796.

Fig. 3.1 The title page of *Pictures of Seventy-Two Beasts and Birds* ..., 1796. Courtesy of the American Antiquarian Society, Worcester, Massachusetts.

states, "He is hunted with dogs and spears, and affords admirable diversion."[24] The parrot is said to "imitate the voice & language of men with more exactness than any other bird."[25] The wolf is described as "a swift, fierce, and voracious beast, destructive to every species of tame cattle; and when prompted by hunger will attack men on horseback, though ever so well armed. They fight in droves, and destroy whatever comes in their way."[26] While some of these volumes remained relatively secular, many tied the study of animals and other wonders of the natural world to Christianity, making them more works of natural theology than natural history. One example of many is the copiously titled, *Fifteen Discourses on the Marvellous Works in Nature: Delivered by a Father to His Children: Calculated to Make Mankind Feel, in Every Thing, the Very Presence of a Supreme Being, and to Influence Their Minds with a Permanent Delight in, and Firm Reliance upon, the Directions of an Almighty, All-good, and All-wise Creator, and Governor* (1791).

Though some were the work of amateurs, most of these natural histories followed the basic format of providing itemized descriptions of plants, animals, birds (which were often placed in their own category), and other natural features and phenomena. A sampling of some printed before 1820 includes *An Easy Introduction to the Knowledge of Nature: Adapted to the Capacities of Children. By Mrs. Trimmer; Revised, Corrected, and Greatly Augmented; and Adapted to the United States of America* (1796) (Figure 3.2); *A Description of the Most Remarkable Birds: For the Entertainment of Children* (1798); *A Short View of the Natural History of the Earth: Designed for the Instrnction [sic] & Amusement of Young Persons* (1803); *Natural History of Birds: Containing a Familiar Survey of the Feathered Creation: To Which is Added, a Short History of Beasts* (1808); *Goldsmith's Natural History: Abridged, for the Use of Schools* (1817); and *Conversations on Natural History: for the Use of Children and Youth* (1820). These natural histories often encouraged children to categorize creatures based upon their degree of utility, classifying them according to their perceived benefit or detriment to mankind. For the most part, the animals are presented as they might be in a museum, posed and provided with only a limited amount of environmental context.

One notable American natural history of this period (and one that was overtly pitched for use in the classroom) was Noah Webster's *History of Animals; Being the Fourth Volume of Elements of Useful Knowledge: For the Use of Schools, and Young Persons of Both Sexes* (1812). In what was probably a cost-saving measure, the book contains no illustrations, but interestingly, it opens with a lengthy section on "Man." It goes into detail about mankind's stages of maturation, general anatomy, and what Webster referred to as the "six distinct varieties, or races" of mankind.[27] It was predictably racist in its delineation of humans; the book refers to

---

24 *Pictures of Seventy-Two Beasts and Birds*, 57.

25 *Pictures of Seventy-Two Beasts and Birds*, 69.

26 *Pictures of Seventy-Two Beasts and Birds*, 17.

27 Webster, *History of Animals*, 16.

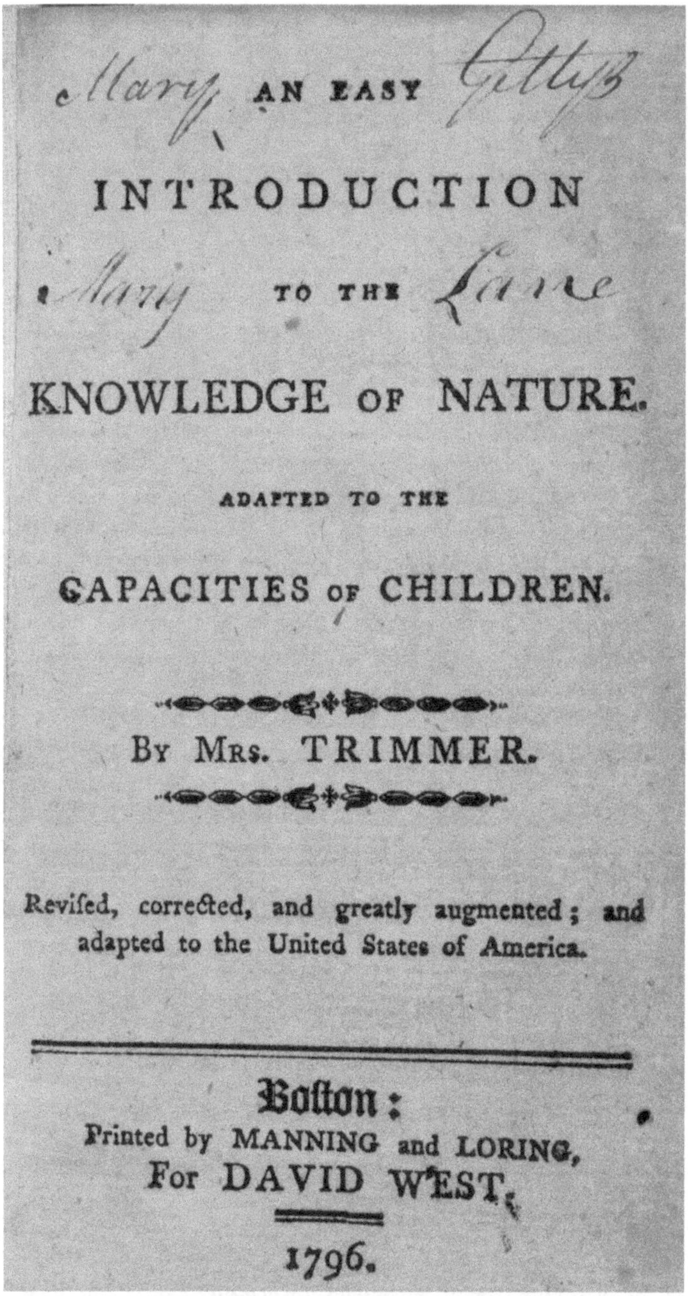

AN EASY

INTRODUCTION

TO THE

KNOWLEDGE OF NATURE.

ADAPTED TO THE

CAPACITIES OF CHILDREN.

BY MRS. TRIMMER.

Revifed, corrected, and greatly augmented; and adapted to the United States of America.

Boston:
Printed by MANNING and LORING,
For DAVID WEST.

1796.

Fig. 3.2 The title page of *An Easy Introduction to the Knowledge of Nature* ..., 1796. Courtesy of the American Antiquarian Society, Worcester, Massachusetts.

Europeans as "the most distinguished race of men" and "Natives of America" as possessing "active bodies and ingenious minds, with the virtues and vices incident to the savage state."[28] It also forwarded the standard Christian worldview typical for the time (it even references Adam and Eve in explaining racial differences). According to Webster, unlike the animals, mankind boasts "a superior degree of intellectual power" and above all "a rational soul, which is capable of continual improvement and high intellectual enjoyment in this life, and is destined to survive to body, and enjoy immortality."[29] Despite these common traits, Webster's domestic natural history was probably a refreshing change from the British reprints. While it did not exclude animals from other parts of the globe it makes a point to include a healthy representation of American animals and anecdotes for its intended American readership.[30]

Juvenile magazines also incorporated regular natural history sections, which were often reprinted extracts from natural histories meant for adults. For example, as early as 1802 *The Juvenile Magazine, or Miscellaneous Repository of Useful Information*, printed in Philadelphia, ran pieces devoted to natural history, promising in its first issue that "It is intended to continue this subject, in the future numbers of the Juvenile Magazine, so as to give a concise system of Natural History."[31] Its first number includes a quotation from Maria and Richard Lovell Edgeworth's *Practical Education*, stating, "Natural History is a study particularly suited to children: it cultivates their talents for observation, applies to objects within their reach, and to objects which are every day interesting to them."[32] The editors followed this epigraph with lengthy reprints from previously published natural history texts, including William Mavor's *Natural History for the Use of Schools*, Thomas Jefferson's description of the mammoth from his *Notes on the State of Virginia*, and "A Short Account of the Lakes ... of the United States" from William Robertson's *The History of America*. Such reprints were common in other juvenile magazines of the time as well.[33]

Years later, in 1828, a different periodical also called *The Juvenile Magazine*, but published in Utica, New York, included its own assortment of natural histories on various animals. For example, between March and September 1828 the

28 Webster, *History of Animals*, 19–20.

29 Webster, *History of Animals*, 10.

30 Webster, *History of Animals*, 15.

31 *The Juvenile Magazine, or Miscellaneous Repository of Useful Information* 1.2 (1802): 99.

32 *The Juvenile Magazine, or Miscellaneous Repository of Useful Information* 1.2 (1802): 99.

33 *The Juvenile Magazine, or Miscellaneous Repository of Useful Information* 1.2 (1802): 99–105, 109–14, and 115–21. For a look at the presentation of natural history to children in the last decades of the nineteenth century and beyond, see Adkins, "'Foundation-Stones': Natural History for Children in St. Nicholas Magazine," 31–47, and Stahl, "Natural History for Children and the Agassiz Association," 213–26.

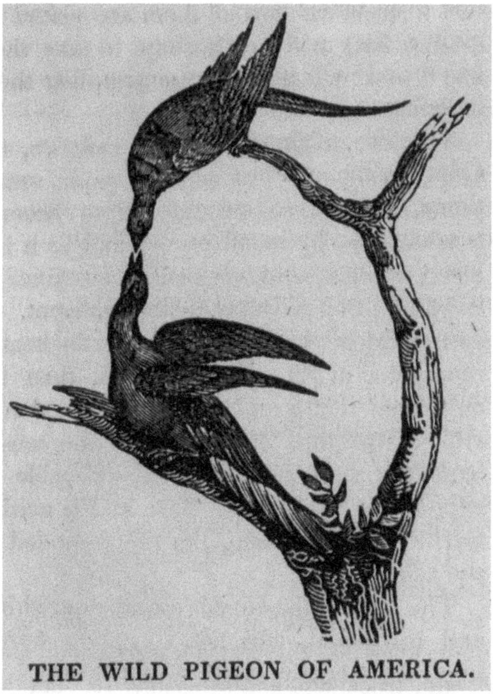

Fig. 3.3 "The Wild Pigeon of America," from *Parley's Magazine*, 1833. Courtesy of the American Antiquarian Society, Worcester, Massachusetts.

magazine included short articles on "Leviathan" (vol. 2, no. 3), "Eagle" (vol. 2, no.4) "Spider" (vol. 2, no. 8) and "Wolf" (vol. 2, no. 9). With the first publication in 1833 of *Parley's Magazine* (published by Samuel Griswold Goodrich), the editors open with a "Prospectus" announcing that among other subjects the magazine would devote a portion of most issues to "Natural History, as birds, beasts, fishes, &c.; as well as plants, trees, flowers, &c.,"[34] and they instantly made good on this promise by including—among an assemblage of poems, mild anthropological vignettes, and pieces devoted to temperance—articles on "The Olive Tree," "The White Headed or Bald Eagle," "The Electric Eel," "The Ass, and "The Wild Pigeon of America" (Figure 3.3).[35]

Most of these were likewise extracts from other publications. The ornithological pieces and their attending illustrations, for example, were mostly reprints from *A Book of Ornithology: For Youth: Embracing Descriptions of the Most Interesting and Remarkable Birds in All Countries, with Particular Notices of American Birds*

34 *Parley's Magazine* 1.1 (1833): 2.

35 *Parley's Magazine* 1.1 (1833): 12, 14, 15, 16, and 94.

(1832), and there are frequent mentions of Audubon's work as well. In each of these three juvenile magazines there is the clear indication that these reprints are predominantly cheap filler for a publication suffering from a lack of content, but it is still telling that the editors chose natural history as that filler. There is a strong sense that natural history is considered a safe and wholesome subject, and one that will help rather than hinder subscriptions.

Juvenile magazines followed the lead of several juvenile books in actively promoting the humane treatment of animals. The idea that children should treat animals with kindness and not engage in cruelty may seem anachronistic, but narratives condemning the mistreatment of animals were common in the juvenile literature of the day, and examples may be found in American juvenile books and magazines from the 1790s forward.[36] Perhaps driven by English reprints such as Sarah Trimmer's *Fabulous Histories: Designed for the Amusement & Instruction of Young Persons* (1794)—a book that promotes kindness to a family of robins—and *The Hare; or, Hunting Incompatible with Humanity* (1802)—a book narrated by the hare himself—American authors eventually joined in as well, with books such as Samuel Goodrich's story, "The Robin's Nest" in *Peter Parley's Juvenile Tales* (1830). In its 1802 review of the Philadelphia reprint of Edward Augustus Kendall's animal welfare novella, *The Sparrow*, the editor of *The Juvenile Magazine, or Miscellaneous Repository of Useful Information* writes, "No part of education is more important than that which teaches children to treat the inferior orders of animals with tenderness and sympathy; for they who after a single admonition, take pleasure in torturing the most contemptible insect, will as they grow up, be cruel and unfeeling to their own species."[37] This reason for teaching children to treat animals with kindness shows the extent to which it is actually based in the desire to promote social welfare, a rationale elegantly linking human well-being with that of the nonhuman. In her study of nineteenth-century botanical children's literature, Maude Hines writes, "[T]his literature that advocated treating the natural world gently was written in tandem with increasing devastation of the natural world—a kind of compensatory romanticism."[38] As Hines suggests, there is something of an inherent hypocrisy in these stories, which insist that children be kind to the natural world while they, in their day-to-day lives, witnessed all manner of environmental degradation and outright cruelty.

Some writers who promoted the humane treatment of animals owned up to the fact that there were definite limits to this kindness. Sarah Trimmer's character of

---

[36] Of course, animal welfare narratives would continue to evolve and proliferate in the years beyond the scope of this study; see Menefee, "Good Masters: Child-Animal Relationships in the Writings of Mark Twain and G. Stanley Hall," 227–44.

[37] *The Juvenile Magazine, or Miscellaneous Repository of Useful Information* 1.3 (1802): 212–4.

[38] Hines, "'He Made *Us* Very Much Like the Flowers:' Human/Nature in Nineteenth-Century Anglo-American Children's Literature," 28.

Mrs. Benson in *Fabulous Histories* praises her children's propensity to treat robins with care, but warns them not to do so at the expense of humanity:

> I am delighted, my dear children, with your humane behaviour towards the animal creation, and wish by all means to encourage it. But though a most commendable propensity, it requires regulation; let me therefore recommend to you, not to suffer it to gain upon you to such a degree, as to make you unhappy or forgetful of those, who have a superior claim to your attention: I mean poor people; always keep in mind the distresses which they endure, and on no account waste any kind of food, nor give to inferior animals what is designed for mankind.[39]

In this thinking the needs of animals are predictably placed below those of the poor, but more interesting is the idea that humans should only be kind to animals to the extent that it does not make a person "unhappy." This condition seems to promote the humane treatment of animals as more of a minor philanthropic luxury, best performed in moderation and never at the expense of another human's well-being. The story in question deals with the act of children giving bread to some robins, but Trimmer's lesson might be applied to larger contexts, such as the destruction of habitat and the mass slaughter of bison, passenger pigeons, and any number of other environmental abuses all performed in the name of making people happy, especially those who stood much to gain economically. Such caveats are absent from other animal cruelty narratives; a notable one, simply entitled, "Cruelty to Animals," in the May 1, 1828 issue of *Juvenile Magazine* recounts a story of a boy who tortured a dog receiving fifty lashes and being forced to wear a sign saying "A savage and inhuman young man."[40] Birds are probably the most common animals portrayed in these cruelty narratives, and one is found in the September 1, 1828, issue of *Juvenile Magazine*, which includes a poem called "On a Goldfinch, Starved to Death in Its Cage," which is narrated by the deceased bird itself, who says it pined for its days in the wild but finally found freedom in death, and, in a rather spiteful moment says to its former owner: "Thanks, gentle swain, for all my woes, / And thanks for this effectual close ... More cruelty could none express; / And I, if you had shown me less, / Had been your pris'ner still."[41]

Apparently, not all animals were worthy of humane treatment or consideration. Immediately preceding "On a Goldfinch, Starved to Death in Its Cage" is an article of natural history on "The Wolf" proclaiming:

> There is nothing valuable about Wolves, except their skins ... They are such ferocious and useless creatures that all other animals detest them: yea, they even hate each other, and therefore, scarcely ever live together, but each one has his separate hole. They are altogether offensive, having a savage look, a frightful

39 Trimmer, *Fabulous Histories*, 16.

40 *Juvenile Magazine* 2.5 (1828): 74–75.

41 *Juvenile Magazine* 2.9 (1828): 141–4.

> howl, an insupportable smell, a perverse disposition, and brutal manners. Perhaps of all other animals, Wolves are the most hateful while living, and most useless when dead ... The continual agitations of this restless animal render him so furious, that he frequently end his life in madness.[42]

Such characterizations of wolves, panthers, and other predators as "altogether offensive" form part of America's lengthy and misguided tradition of exterminating top predators from its landscape. That these descriptions are so cartoonish and often patently false demonstrates how they, like similarly vicious portrayals of Native Americans, were part of a larger cultural project of removal and extermination and served likewise as a rationalization for that very policy. These kinds of descriptions helped justify the animal's extermination as well as rationalize it afterward. Indeed, reading that a wolf is so perverse a character that "he frequently end his life in madness" makes the extermination of such "useless creatures" sound like a humane act.

Wandering the Woods

Following the Revolutionary War there was a simultaneous increase in the number of British reprints as well as in the number of original works by American authors. Many tales that had been popular for decades also hung on in new editions. A few of these early American juvenile narratives incorporated the wilderness as an important, if still one-dimensional, component of the plot. These stories often deal with children entering the woods, where they variously perish from exposure, are rescued in the nick of time, or enjoy a pleasantly educational stroll through the trees. Probably the oldest and most popular of these tales is *The Affecting History of the Children in the Wood*, which has its origins in the popular sixteenth-century tale of *The Babes in the Wood.* In America this tale went through dozens of editions, in verse and in prose, in English and German, between 1760 and 1814. The popularity of the story is evidenced by the various iterations of its title, which was printed as *The Babes in the Wood, The History of the Children in the Woods, The History of the Two Babes in the Wood*, and so on. In this well-known tale a wealthy man and woman die, leaving their two children in the care of an evil uncle who plots to kill them to inherit their family fortune. In a 1796 version entitled, *The Affecting History of the Babes in the Wood*, the story follows its usual course when the children's uncle hires "two ruffians to carry them to a distant wood, where he proposed that they should be murdered."[43] Once in the woods one ruffian has second thoughts and kills his accomplice to save the children's lives. He then leads them deep into the woods, leaving them there with the promise that he will soon return with bread. Afraid to wander too far from that spot, the children stave

---

42 *Juvenile Magazine* 2.9 (1828): 142.

43 *The Affecting History of the Babes in the Wood*, 10–11.

off their hunger with blackberries before falling asleep on the forest floor. As the story goes,

> The severity of the night, joined to their want of food, and grief of heart, were more than their tender frames could bear, and they expired in the midst of this unfrequented wood. Their delicate bodies were exposed to the open air for several days, till at length two Robin-Red-Breasts, touched as it were with human sympathy, conveyed leaves to the place where the children lay, and covered them with the tenderness of the most affectionate parents.[44]

The actions of the birds are attributed to "that kind Father who never fails to administer comfort to the sons and daughters of affliction."[45] After a series of personal tragedies and becoming confined to debtor's prison, the evil uncle confesses his crime and is executed. While the story is clearly a fable meant to both warn children against the evils of greed and to exhort them to appreciate their parents, it does bear some interesting environmental elements. The wood in itself is not an evil place; it is not even populated with dangerous wild animals. Rather, the children die of exposure and grief and actually receive belated care from the birds, who act both as surrogate parents and as the hand of God.

The story of the Babes in the Wood inspired variations, such as the British author Richard Johnson's (1733–1793) *The Hermit of the Forest, and the Wandering Infants: A Rural Fragment*, printed in 1789 by the Boston printer Samuel Hall. In this English tale two children are abandoned in the woods by their evil uncle and arrive on the doorstep of a kind hermit who has taken refuge in the woods to escape the abuses of the Scots. The hermit cares for the children, telling them stories, and teaching them to read. He is, however, loath to return them to civilization because he fears it will give away the location of his hut, and he warns the children, "[S]hould you wander far from this hut, you may lose yourselves in these woods, and no more find your way home, when you will die with hunger and thirst."[46] The children stay with the hermit and are, in a further nod to the Babes in the Wood, befriended by birds, "who were so tame as to fly round them, and would sometimes even suffer themselves to be stroked by them."[47] The story is resolved when a group of foxhunters arrive (the fox also takes refuge in the hermit's hut), and in true fairytale fashion, among their hunting party is none other than the children's father. Unlike its predecessor, *The Hermit of the Forest* has a happy ending, a trait that became increasingly common in the various walk-in-the-woods-style books as the late eighteenth century gave way to the early nineteenth century. One good example of a story relating a non-lethal encounter with the wilderness is Isaiah Thomas's 1794 Worcester, Massachusetts, edition of *The Mountain Piper; or, The History of Edgar and Matilda*. This reprint

---

44 *The Affecting History of the Babes in the Wood*, 15–17.

45 *The Affecting History of the Babes in the Wood*, 17.

46 *The Hermit of the Forest*, 18.

47 *The Hermit of the Forest*, 18.

of an English pastoral story tells the tale of a gentleman who finds delight in rural strolls and the simple ways of the lower classes who live in the rustic region. It is filled with lush descriptions of the "natural beauties" of the "romantic country."[48]

An intriguing American book about a rural stroll, and one unencumbered by abstract pastoralism, is *A Walk and Conversation between a Fond Father and His Little Son: As They Took a Walk through the Fields and Meadows, &c.: For the Use of Children*, printed by Trumbull in Boston in 1804. Presented as a dialog between a father and son on proper agricultural practices, this slim book includes lessons on moral behavior and follows the two as they, in the words of the father, "walk through the fields, the pastures, orchards, and meadows, and perhaps through the wild woods, so as to contemplate the works of the God of Nature, as well as the works of Man."[49] The conversation that ensues touches on an impressively sophisticated range of early American agrarian topics from the cultivation of apples to the evils of drinking too much cider; from the winter care of horses and oxen to the growing of corn and pumpkins. Their discussion even touches on how to build bridges and on the way that the shipping of America's agricultural goods ("Pork, Beef, Potash, Flaxseed, Flour, and Lumber") overseas allows Americans to pay for the acquisition of "Linnen and Woolen Cloths, &c. from Europe, Tea, Sugar, and China, from the East Indies; and Rum, Molasses, and Sugar &c. from the West Indies."[50] Central to this text is the understanding that American national well-being is firmly tied to the land. The father makes much of how the rivers drive mills that allow for the processing of corn and wheat into flour while also reducing America's dependence on Europe for the production of paper. Following this look at America's fledgling industrial efforts, the two make their way to a waterfall, where the text shifts to an appreciation of the nation's natural beauty. The father describes the cataract in language that mingles images of abundance with the sublime:

> After the river has run a great ways winding about the vallies, it comes to a heap of rocks and stones, and there being a hill each side, it is obliged to tumble over the rocks and it falls down a great way among the stons [*sic*], and makes the water foam and [turn?] white as milk, and boil like a pot over [a?] fire, dont you hear the noise? dont you see how the spray of the water flies up and looks almost like rain? sometimes when the sun shines upon it there will appear several rain bows, somewhat like that you see a little before night when it rains and the sun shines at the same time.[51]

The father follows this majestic scene with a description of Niagara Falls, and then, in tracing the water's course to the ocean, he describes the nation's whaling industry. As the two head home for dinner, the father explains the great benefit of using taxes to maintain public roads, and says that the road they travel "used

48 *The Mountain Piper*, 7, 9.

49 *A Walk and Conversation*, 5.

50 *A Walk and Conversation*, 18.

51 *A Walk and Conversation*, 25–6.

to be called the Kings, but of late years we do not allow the king of England to be our King, so now we call it the public road, or what belongs to the Republic of the United States of America."[52] *A Walk and Conversation* provides a detailed look at the American agrarian lifestyle at the start of the nineteenth century, and with the glut of foreign reprints lining their bookshelves, young Americans probably enjoyed reading a book that was about, and for, Americans. Its portrayal of the natural world, mixed with its praise of republican virtues and agricultural prowess, is refreshingly straightforward and in many ways stands as a precursor to the kind of proto-environmental histories Samuel Griswold Goodrich would publish 20 years later.

As the years progressed, more juvenile books by American authors appeared, and a number of them took facets of the local environment as the subject matter. One such book is *The Hunters, or The Sufferings of Hugh and Francis, in the Wilderness: A True Story*. Authored by the New England textbook writer Caleb Bingham, and printed by Samuel Armstrong in Boston in 1814, this book relates the brief and ostensibly true tale of a 1778 moose hunting trip taken by two Dartmouth students into what Bingham describes as "the then thick and gloomy forests of Vermont."[53] Equipped with snowshoes, the two boys—Hugh, a merchant's son, and the other boy a mixed-race Native American named Francis (a member of the St. François tribe)—enter the woods, where the snow is said to be five feet deep. After three days of hunting, Francis kills a moose and saves them from starvation, but soon thereafter Hugh falls and accidentally stabs himself with his hunting knife. Francis cares for this wounds and helps fend off wolves. With their provisions low, Francis goes in search of game and comes across what Bingham calls a "flock of moose" where he kills three of the snow-encumbered animals with his hatchet.[54] Bingham elaborates on the nature of the moose, providing a detailed explanation of their habits. He writes, "It is observable that these animals, during a great depth of snow, herd together in a grove of spruce, which they live upon for the time being, treading the snow hard under their feet, like neat cattle in a farmer's barn yard. I have known thirteen killed, in such a situation, within a few rods of each other."[55] Applying his genius for wilderness survival, Francis cares for the injured Hugh for an entire month, all while taking care not to work on Sunday (as he was warned against by the president of Dartmouth before leaving on their trip). Eventually the two make their way out of the wilderness (Francis carries Hugh on his back much of the way) and return to Dartmouth two months after they left. Holding up the heroic Francis as his example, Bingham concludes the story by admonishing "those who are accustomed to depreciate the character of Indians, and who are more disposed to extirpate them from the face of the earth,

52 *A Walk and Conversation*, 28–9.

53 Bingham, *The Hunters*, 10.

54 Bingham, *The Hunters*, 20.

55 Bingham, *The Hunters*, 20.

than to civilize and Christianize them."[56] Looking beyond Bingham's argument for the assimilation of Native Americans to his portrayal of the environment itself, he presents his young readers with a Vermont wilderness that while dangerous, is not itself a place of evil. Bingham does, however, forward the wilderness as a place where shelf-worn stereotypes predominate: in the wilderness the white Hugh proves useless to the point that he is a hazard to himself and others, and Francis, the Indian, proves heroic. This idea of the American wilderness as a proving ground of masculine capability was hardly new in 1814, but it would, with the post-Cooper rise of the Frontier romance, become an even more pronounced element of American literary culture.

Samuel Goodrich's Formation of a Juvenile Environmental Literature

By the 1820s, juvenile books began to provide a more sophisticated portrayal of animals, their environments, and the human impact upon the land, and nowhere was this combination of elements as pronounced as in the works of Samuel Griswold Goodrich (1793–1860), better known to millions of children as Peter Parley. In his early study of American juvenile literature, A.S.W. Rosenbach calls Goodrich's invention of Parley "one of the most momentous and influential events in the history of American children's literature in the nineteenth century."[57] Goodrich might also be viewed as the inventor of a new brand of American juvenile literature, one that consciously moved the natural world to the foreground while also linking it with the push to create a distinct American literary culture.

When, in 1856, he published his autobiography *Recollections of a Lifetime, or Men and Things I Have Seen*, Goodrich could look back on decades of success in a variety of occupations. Active in politics, he had been elected to the Massachusetts House of Representatives in 1836, and in the following year he was elected to the State Senate. Ever the cosmopolitan gentleman, Goodrich was a natural choice for the position of U.S. consul to Paris, a role he held from 1851 to 1853. Yet it was Goodrich's head for the publishing business that secured his fortune and fame. He had an eye for talent and a knack for marketing his books to his best advantage. He was also interested in fostering what was at that time a very tentative domestic literary movement. In the late 1810s he set about to promote (and hopefully profit from) the work of American authors in a marketplace dominated by British reprints.[58] In his autobiography he writes,

> I began to think of trying to bring out original American works. It must be remembered that I am speaking of a period prior to 1820. At that date, Bryant, Irving, and Cooper—the founders of our modern literature—a trinity of genius in poetry, essay, and romance—had but just commenced their literary career.[59]

---

56 Bingham, *The Hunters*, 33–4.

57 Rosenbach, *Early American Children's Books*, xlviii.

58 This biographical information is drawn from Daniel Roselle's *Samuel Goodrich, Creator of Peter Parley*.

59 Goodrich, *Recollections of a Lifetime*, 2.110.

One area of publishing that especially interested Goodrich was juvenile literature. He writes, "I turned my attention to books for education and books for children, being strongly impressed with the idea that there was here a large field for improvement."[60] In working toward that goal, Goodrich succeeded not only in promoting American books to American children but he also, in attempting to teach young Americans about their country and themselves, managed to create a series of books with an environmental awareness often lacking from popular literature at the time. Attempting to create a viable American form of juvenile literature, Goodrich produced a handful of books with a surprisingly mature, if ultimately ambiguous, environmental message, and he would do it through one of the most unlikely of means, with his creation of a fictional old Bostonian named Peter Parley.

Born and raised on a small farm in rural Ridgefield, Connecticut, Goodrich attended school infrequently, but enough that he had the basics of literacy under his belt by the age of 10. His father was a minister, and key to his further education was his father's library, where he read the *New England Primer* (which he recollected were "embellished with hideous cuts of Adam's Fall, in which 'we sinned all'")[61] and Watt's *Divine and Moral Songs*; he also perused a selection of other books on his father's shelves, most of which were for religious instruction.[62] Goodrich biographer Daniel Roselle writes that the young man's "first encounters with Mother Goose were 'revolting,' 'shocking,' and 'deeply painful' to him—so much so, in fact that he resolved to combat the Gooseian incantations that cows jumped over the moon instead of giving milk."[63] Goodrich, perhaps taking some pride in how much he personally helped increase and enrich the selection of juvenile books in America, writes, "It is difficult now, in this era of literary affluence ... to conceive of the poverty of books suited to children in the days of which I write."[64] After reading the Mother Goose rhymes "with no real relish," young Goodrich borrowed a volume of tales that included the stories of Little Red Riding Hood, Puss in Boots, Blue Beard, and Jack the Giant-killer, each of which he refers to as "tales of horror, commonly put into the hands of youth, as if for the express purpose of reconciling them to vice and crime."[65] His response to these and other tales is especially illuminating, as it reveals his innate belief that juvenile literature should educate children in how to live their lives, and in so doing should not lead children astray, either from basic material reality or from developing a sense of propriety. For Goodrich,

---

60 Goodrich, *Recollections of a Lifetime*, 2.112.

61 Goodrich, *Recollections of a Lifetime*, 1.165.

62 Roselle, *Samuel Goodrich, Creator of Peter Parley*, 13–14.

63 Roselle, *Samuel Goodrich, Creator of Peter Parley*, 15.

64 Goodrich, *Recollections of a Lifetime*, 1.165.

65 Goodrich, *Recollections of a Lifetime*, 1.166.

children's books were either full of nonsense, like "hie diddle diddle" in Mother Goose, or full of something very like lies, and those very shocking to the mind, like Little Red Riding Hood. ... How far this supper of horrors familiarized my own mind with violence, and thus defaced that moral sense, which is common in children—leading them to prefer the good, the true, and the beautiful, if it be duly cherished—I cannot venture to say."[66]

Despite his dislike of his early reading materials, in another, perhaps even more formative area of his early education—the appreciation of the natural world—the young man could not complain. In recalling his rural upbringing he writes,

In my youth I became familiar with every bird common to the country: I knew his call, his song, his hue, his food, his habits; in short, his natural history. I could detect him by his flight, as far as the eye could reach. I knew all the quadrupeds—wild as well as tame. I was acquainted with almost every tree, shrub, bush, and flower, indigenous to the country; not botanically, but according to popular ideas. I recognized them instantly, wherever I saw them; I knew their forms, hues, leaves, blossoms, and fruit. I could tell their characteristics, their uses, the legends and traditions that belonged to them. All this I learned by familiarity with these objects; meeting with them in all my walks and rambles, and taking note of them with the emphasis and vigor of early experience and observation.[67]

As this passage suggests, while the juvenile literature of Goodrich's day shocked his sensibilities, he found the natural world far more suited to his tastes. He asserts, "[T]his communion with nature gave me instructive and important lessons from nature's open book of knowledge."[68] This admiration of the natural world would make its way into many of his books, from his juvenile fiction to his textbooks on geology, botany, and natural history.

Around 1819, Goodrich began to publish books with an American juvenile audience in mind. For Goodrich that meant writing and promoting books that were overtly about United States history, people, and environment. This combination of subjects in American juvenile literature, at the nationalistically crucial literary moment of the 1820s, demonstrates a key step in the development of a more sophisticated and regionally distinct American environmental consciousness. This stage in the early evolution of America's environmental literature is particularly significant because it occurs in those very works of literature designed to influence young minds and therefore positioned to impact the nation for decades to come.

Though his name has since become obscure, Goodrich's books were extremely popular in his day. His fictional narrator, Peter Parley, was famous on both sides of the Atlantic as a kindly Bostonian grandfather figure with a cane, a bum leg, and a wealth of stories to share. Between 1827 and the 1840s Parley narrated (or had his name lent to) dozens of Goodrich's juvenile books, and—in the absence

---

66 Goodrich, *Recollections of a Lifetime*, 1.167–8.

67 Goodrich, *Recollections of a Lifetime*, 1.104–5.

68 Goodrich, *Recollections of a Lifetime*, 1.103.

of copyright law—there were even more pirated and copycat books bearing the Parley name. In his autobiography Goodrich writes, "I thus stand before the public as the author and editor of about one hundred and seventy volumes—one hundred and sixteen bearing the name of Peter Parley. Of all these, about seven millions of volumes have been sold: about three hundred thousand volumes are now sold annually"[69] With this large readership and with his portrayal of an American environment that does much more than serve as window dressing, Goodrich stands as one of the most important figures in the formation of an early American environmental literature. He admitted in his autobiography that while he saw room for improvement in juvenile literature, he began his career without much of a plan:

> It is quite true that when I wrote the first half-dozen of Parley's Tales, I had formed no philosophy upon the subject. I simply used my experience with children in addressing them. I followed no models, I put on no harness of the schools, I pored over no learned examples. I imagined myself on the floor with a group of boys and girls, and I wrote to them as I would have spoken to them.[70]

This approach probably worked to Goodrich's favor because, by ignoring any model except that of speaking directly to children, he created an engaging literary style, and touched on subjects in ways that had not been fully realized. While authors of juvenile literature had used narrative personae and first-person narrators before, few if any had done so with the warmth and charm found in Goodrich's books. In further explaining his approach to teaching children, he gives much credit to his understanding of "all children to be ardent lovers of things they could see and hear and feel and know"[71] a fact he capitalized on by including pictures in his books and by commented directly upon them. He writes,

> Thus I sought to teach them history and biography and geography, and all in the way in which nature would teach them—that is, by a large use of the senses, and especially by the eye ... I selected as subjects for my books, things capable of sensible representation, such as familiar animals, birds, trees, and of these I gave pictures, as a starting point.[72]

Through these methods, Samuel Goodrich managed to promote to America's youth the humane treatment of animals, the understanding of American natural history, the appreciation of the beauty of nature in general, and the development of upright moral character, and he did it all in an ambitious string of juvenile books that, while they were not called such at the time, were in fact progressive works of environmental fiction, environmental history, and social criticism.

---

69 Goodrich, *Recollections of a Lifetime*, 2.543.

70 Goodrich, *Recollections of a Lifetime*, 2.308.

71 Goodrich, *Recollections of a Lifetime*, 2.310.

72 Goodrich, *Recollections of a Lifetime*, 2.310–311.

Goodrich's earliest attempts at writing for children are not akin to anything we would see on children's bookshelves today, and in fact, they are also very unlike Goodrich's later Peter Parley tales. They do, however, often take animals as their subjects, and in ways that differed from other books of the time. In his autobiography he provides a brief account of how he began his career as a writer and publisher of juvenile literature, relating that in 1819

> I turned my attention to books for education and books for children, being strongly impressed with the idea that there was here a large field for improvement. I wrote, myself, a small arithmetic, and a half a dozen toy-books, and published them, though I have never before confessed their authorship. I also employed several persons to write school histories, and educational manuals of chemistry, natural philosophy, &c., upon plans which I prescribed—all of which I published; but none of these were very successful at that time. Some of them, passing into other hands, are now among the most popular and profitable school-books in the country.[73]

Goodrich made a push into several areas of juvenile book publication, from schoolbooks to what he calls his early "toy-books." At least five of these tiny story books remain extant, and these slim volumes each bear the secondary title of *A New Tale for Children*. This handful of early stories includes *The Two Doves and the Owl*; *The Man and the Snake*; *The History of a Little Silver-Fish*; *The Vagabond*; and *A Story of Little Thomas and Betsey*. Each of these small, 16-page chapbooks appeared while Goodrich was still conducting his publishing business from Hartford, Connecticut. In fact, they are closer in nature to the disturbing stories of his youth (that "supper of horrors") than to the more upbeat historical tales that would eventually make him so successful. Each of these tiny books reveals a writer who is experimenting with subject matter and voice, who is trying to create a juvenile literature that is profitable and instructive. In these earliest of his tales, we see a writer getting his bearings by creating stories that largely match the content and presentation found in other juvenile tales. The result is a handful of stories that represent a moment of literary throat clearing that preceded Goodrich's invention of Peter Parley and his signature style.

Goodrich's *The Two Doves and the Owl*, for example, illustrates how he borrowed the moral severity of earlier juvenile literature while still making good on his desire to avoid any overly fantastic portrayal of animals. Goodrich describes the two doves as morally upright creatures: "They were very beautiful birds—they had bright shining feathers, and sweet melancholy voices. They were very harmless too—they did not devour worms, but lived upon grains which they gleaned from the fields."[74] The two birds soon have a nest filled with "four pretty young ones," and one day while the father dove is gone, "a great Owl came along

---

73 Goodrich, *Recollections of a Lifetime*, 2.112.

74 Goodrich, *The Two Doves and the Owl*, 4.

and attacked the dove which was sitting on the nest."[75] After a battle, the female dove is killed and the "naughty owl," whom Goodrich writes, looks like "a sour old batchelor," eats the nestlings.[76] The father dove returns to the bloody scene where he finds "his little ones ... gone, his pretty mate ... dead upon the ground," and, with a tragic finality unknown to our modern children's literature, the story ends with the male bird's attempt to rouse his mate: "But it was all in vain! The little dove was dead—and her eyes were shut for ever."[77] As is the case with many of these early chapbooks, *The Two Doves and the Owl* is actually just the first of three short prose pieces in the book. Following the account of the doves is a tale about how a pet antelope is employed to send an important message to a little princess in danger of kidnapping, and Goodrich concludes the book with a brief description of the fox, which he claims as "the most sly and cunning of animals," with "a careful designing face," making him "a very disagreeable fellow."[78] Goodrich uses this description of the fox (who "skulks about in the night and carries off geese and fowls which he tears in pieces and devours") as a cautionary figure, advising children to "*never be cunning*—for cunning folks are always found out, and then all confidence is withdrawn by society from them."[79]

In these examples of some of Goodrich's earliest short stories, he employs animals in relatively traditional ways for juvenile literature of the time, but his description of the natural habits of the animals are, for the most part, grounded in reality. His creatures stand as moralistically extreme emblems of virtue or sin, and in each case they instruct children about proper conduct in, as well as the dangers of, the human world. Serving as proxies for humans, the doves, owls, antelope, and foxes of Goodrich's early books become charismatic placeholders for parents, friends, siblings, and the children themselves. As he would throughout his career, Goodrich reveals a far greater concern with the integrity of the family unit than with actual animals. The story of the mate-killing, nestling-eating "old batchelor" owl serves as a lesson in the value of family and for the need of family members to ensure each other's safety, a lesson evidenced by the dove father's absence precipitating the demise of his entire family. Goodrich pairs this domestic theme with a more innovative tendency to portray the animals as actual animals. While they do shoulder a good amount of anthropomorphism, they never act too much beyond the range of natural possibility, and this strategy makes sense when placed next to Goodrich's distaste for *Mother Goose's Melodies* and *Little Red Riding Hood.*

In another of his 1819 chapbooks, *The Man and the Snake*, Goodrich criticizes the well-worn myth that rattlesnakes have the ability to hypnotize their prey, and even humans. "Come my little reader," writes Goodrich at the opening of the story,

---

[75] Goodrich, *The Two Doves and the Owl*, 4–5.

[76] Goodrich, *The Two Doves and the Owl*, 5.

[77] Goodrich, *The Two Doves and the Owl*, 6.

[78] Goodrich, *The Two Doves and the Owl*, 14.

[79] Goodrich, *The Two Doves and the Owl*, 7–13, 15.

> did you ever see a snake? Snakes are very disagreeable animals, though all of them are not hurtful. ... They are generally harmless creatures, and are very shy unless injured or provoked. Some of them are poisonous, others have the power, it is said, of *charming* birds, squirrels, and sometimes men.[80]

The story then progresses as a frame tale in which a man visits the narrator to relate how after a nap in the woods he woke to find a snake nearby.

> As soon as I fixed my eyes upon him, his bright yellow spots grew still more brilliant—his eyes assumed the appearance of small dots of flame, and the rattling subsided into a soft and pleasant humming. ... I thought I saw the most beautiful colours, and the most charming scenes. I no longer saw the snake, but thought myself in some new and happier world.[81]

The man concludes his tale by saying that after the snake coiled around his legs that he woke from his trance and escaped. However, Goodrich's narrator (a prototypical Peter Parley) will have none of it, and ends the narrative with a refreshing candor, saying to his young reader: "But I hardly believe this story true, for I have seen a great many snakes, and I believe if you let them alone they will let you alone."[82] And he then shifts to the next story by—in a fashion common to Goodrich's writing—calling attention to a woodcut that signals the next story in the book: "Well! Here comes a monkey, drinking a glass of wine."[83] The narrator's dismissal of the man's snake story reveals how Goodrich's approach to writing about animals differed from other juvenile books at the time. He presents a folktale about a rattlesnake charming a man only to burst the irrationality of those kinds of tales, and he does it with a sensible authority: he has "seen a great many snakes" and never witnessed anything resembling the man's tale, and he ends with the sound advice that in leaving snakes alone they will present no danger. Earlier in the story we get hints of the narrator's down-to-earth sensibilities when he states that snakes "are generally harmless creatures, and are very shy unless injured or provoked." Thus Goodrich's narrator works counter to the fanciful portrayals of animals and deflates a common myth about their abilities to harm people. When contrasted with Crèvecoeur's bogus portrayal of the supernatural qualities of snakes in *Letters from an American Farmer*, it becomes clear that while Crèvecoeur's claims might be uncontested by his European audience, Goodrich is writing for those children who have already, or who doubtlessly will, encounter snakes in the course of their American lives, and hence he replaces a myth with advice for avoiding injury.

In 1827 Goodrich published the first of his Peter Parley books, *The Tales of Peter Parley, About America*. The book is a first-person history of New England presented by the fictional Parley, who regales his child audience by drawing upon

80 Goodrich, *The Man and the Snake*, 4.
81 Goodrich, *The Man and the Snake*, 5–6.
82 Goodrich, *The Man and the Snake*, 6.
83 Goodrich, *The Man and the Snake*, 6.

his own purported adventures—spanning the French and Indian War, the American Revolution, and on up to the present day. This book was a hit, and Goodrich soon followed with more. Not all were listed as books in the Parley series, but virtually all of them borrowed some of the elements of his engaging Parley formula. Interested in using the Parley persona for works of fiction and nonfiction alike, Goodrich wrote a handful of both, and most of these books, of either stripe, contain what at the time must have seemed like radically environmental statements that, while rarely of a kind we would find notable today, at least recognize the reality of human-created environmental degradation and depletion.

Goodrich's *The Child's Botany* (1828) is an intriguing example of how he balanced his recognition of anthropogenic environmental change with a desire to teach New England children about the unique and valuable character of their region's natural world. While not marketed as a Peter Parley book, *The Child's Botany* still bears some of the most characteristic hallmarks of Goodrich's inviting style. His first-person narration speaks directly to his juvenile readers; he is down-to-earth in his explanations; he includes helpful illustrations; and he asks questions that encourage active participation with the text. One representative passage asks children to consider the changes that have occurred to the land since settlement:

> Do you know what great forests there are in America? They were much greater formerly, than they are now. Two or three hundred years ago, all the country about New York, and Boston, and Hartford, was covered with woods. In these the Indians lived; and they were full of bears, wolves, panthers, deers, foxes, and many other animals. But when our forefathers came from England, they cut down the woods, to prepare the land for grain, grass, and fruit trees. For the White people were civilized, and they wished to get their living by farming and not by hunting, as the Indians did. But far to the West, there are now forests, which are a great many miles in extent, and where the Indians and wild beasts now live.[84]

When isolated from its context, this extract might seem subversive, to beg condemnation of the land use practices that caused such rapid and dramatic change to the environment, but that is not the case. Rather, Goodrich takes an approach common in several of his books, both those for children and adults, by presenting facts (at least the facts as he viewed them) without the judgmental commentary that he seems to set up. For example, in this passage he points out that the land of 1828 New England had seen widespread deforestation since the arrival of Europeans, but he only mildly concedes that the denuding of the landscape denotes a widespread problem of environmental exploitation, not to mention the removal of the Native American communities, to fit the European American colonial agenda.

That is not to say that Goodrich fails to, in his own way, call into question the wisdom of this destruction, or at the very least bemoan the loss of diversity

[84] Goodrich, *The Child's Botany*, 14–15.

and natural beauty that followed in its wake. Instead of outright disapproval he instead employs what might be termed a form of environmental nostalgia, one that stops short of condemnation. For instance, he reminds his juvenile readership that, before European settlement, the land that is now their home "was covered with woods. In these the Indians lived; and they were full of bears, wolves, panthers, deers, foxes, and many other animals."[85] In looking backward to this earlier environment, Goodrich invites a comparison between the imagined pristine land of years before and the land as it looked to Goodrich's readers in 1828. He is careful not to overtly judge the practices that have led to the present scene, leaving that task for his readers. In fact, Goodrich is so consistently non-judgmental of what European Americans have done to the natural world, that he appears neutral on the subject. As is the case in several of his books, he seems much more concerned that children learn to enjoy and respect the natural world than to question those land use practices that have caused such wholesale change.

Goodrich's 1830 *Peter Parley's Juvenile Tales*, a work of fiction containing two short stories that each convey an environmental lesson, continues his treatment of issues related to the natural world while also packaging his sentiments in a manner appealing to New England children. The first tale of the book, "The Bird's Nest," is about three curious children who remove a robin's nest from a tree and place it, along with the nestlings, in their house (Figure 3.4).

The parent birds fret and eventually leave the area never to return, and the nestlings die from the children's inadvertent mistreatment of this small avian family. As the boy and his two companions remove the nest from the tree, Goodrich invites us to consider the feelings of the birds:

> Oh! how pitifully did the parents of the little robins cry and scream, round the cruel boy. But he heeded them not. He tore the nest from its place, and handed it to the largest of the girls. She took it, and they carried it, with the four little robins away. The parents of the young robins, followed the children wailing as if their hearts would break. But it was all in vain. The children took the young robins into the house, and fed them, but in a few days the poor creatures all died. ... It is a cruel thing thus wantonly to rob the poor birds of their nests, and I hope none of my little readers will be so hard hearted as to do so. ... none but cruel and unfeeling persons can wish to turn their pleasure into mourning.[86]

Here Goodrich balances storytelling with didactic asides. He creates a strong correlation between the family of birds and the family of people who has in their unthinking appreciation doomed the bird family's existence. Goodrich invites parallels between these two families, between their use (and abuse) of the environment, and between their common love for their offspring. By thoroughly humanizing the birds, he encourages environmental empathy on a level rare not only for juvenile literature of the time but also for works intended for adult readers as well.

[85] Goodrich, *The Child's Botany*, 14.

[86] Goodrich, *Peter Parley's Juvenile Tales*, 12–14.

Fig. 3.4 Frontispiece, "The Bird's Nest" in *Peter Parley's Juvenile Tales*, 1830. Courtesy of the American Antiquarian Society, Worcester, Massachusetts.

The second story in *Peter Parley's Juvenile Tales*, "The Faithful Dog," relates the plight of a boy named Henry, who during a pleasant stroll in the woods injures his leg. His dog, Trusty, stays with him, protecting the boy from wild animals and barking so that rescuers may hear. The narrative tension of this brief story pales in comparison with Goodrich's beautiful descriptions of the forest, and indeed the tale seems to exist for the sake of these descriptions. In relating Henry's journey into the woods, Goodrich writes that "Henry had set out early in the morning, and Trusty went barking and jumping by his side. He was already on the hill, when the sun was rising."[87] Then, in a narrative voice perhaps as much Goodrich as Parley he addresses his young readers in general, proclaiming:

> It is beautiful to be in the country in spring, when the sun comes up! Before the sun yet appears, the robin is on the top of the tree singing his song; the larks, and blackbirds, and chipping birds, and martins, are all awake, and seem to be talking to each other as fast as they can. The grass is sparkling with dew, the vallies are white with mist, and the air is sweet with the smell of flowers. I recommend it to all of my little readers, when they happen to be in the country in spring, to go upon the hills, and see the sun rise.[88]

So while the narrative relates an instance of danger in the wilderness, the descriptions and advice of the narrator make plain that Henry was right to go into the woods and that children should, with safety in mind, also make such ventures in the woods for there they will glimpse unparalleled natural beauties.

Goodrich's advice to children to enjoy the beauty of the natural world despite its occasional dangers is worth noting because it runs counter to any simplistic view of the land as either a threat or a source of beauty and wellness. The woodlands of the Peter Parley tales are idealized, but not excessively so. They are, rather, portrayed as complicated settings in which children learn valuable life lessons. Goodrich, unlike any other writer of fiction in his day except for Cooper with *The Pioneers* (which is clearly an influence on Goodrich's work), offers his readers scenes of the natural world that while romanticized rarely devolve into mere allegorical or fantastic landscapes. He also does not shy away from divulging the ways that human encroachment has altered the land, though he never condemns such degradation in the manner found in Cooper's *The Pioneers*.

As it turns out, a good example of Goodrich's complicated environmental vision occurs in *Peter Parley's Story of Little Marion* (1830). In this brief tale Goodrich essentially retells Cooper's panther attack scene from *The Pioneers* and in so doing re-engages with the environmental ideas that the scene puts into motion. At the opening of Goodrich's tale, young Marion Leslie's parents have moved their family to the wilds of eastern Ohio,[89] a place of intense natural beauty, where their cabin

---

87 Goodrich, *Peter Parley's Juvenile Tales*, 18.

88 Goodrich, *Peter Parley's Juvenile Tales*, 18–19.

89 Aside from the story's debt to James Fenimore Cooper's *The Pioneers*, Goodrich may have lifted Marion's surname from the Leslie family of Catherine Maria Sedgwick's *Hope Leslie* (1827); Marion's unguarded love of nature and the story's Ohio River Valley setting is also reminiscent of Gilbert Imlay's *The Emigrants*.

> was surrounded by thick woods ... and at a short distance the river flowed on so smoothly ... it was, indeed, as beautiful as a scene of wild nature can be. The vast forest trees spreading their thick shade over the land; the luxuriant wild flowers putting forth their gay blossoms, the clear ringing music of the birds that were happy and at home in the broad forest, made it seem quite a paradise.[90]

As he does in virtually all of his other forest stories, Goodrich paints a scene of almost overwhelming natural beauty while also stopping short of outright fantasy. This landscape, as almost supernaturally verdant and delightful as it is, retains a level of plausibility. Goodrich makes it clear that Marion and the Leslie family live a believable life on the frontier, subsistence hunting and scraping by on the edge of the European American colonial blade as it slices from east to west.

Marion is enamored with their locale on the Ohio frontier, saying to her father, "[H]ow much more beautiful is this, than the cold shorn hills of New Hampshire."[91] The story that follows Marion's pronouncement seems calculated to complicate her pastoral sentiment, and to serve as a caution to readers who might likewise oversimplify and underestimate the American wilderness by praising its beauties while remaining blind to its dangers. The story begins on a September day when little Marion decides to go for a walk in the woods to find her brother, Alexander, (who is late in returning from a hunting trip) and to enjoy the landscape's beauty. Goodrich writes,

> It was not unusual for Alexander to go alone into the woods, for though the catamount and cougar and other beasts of prey were occasionally met with in the woods, yet they had never been seen near Mr Leslies's house; and besides, they would seldom attack a man and were therefore little feared or thought of."[92]

Intending to walk only a short distance in search of her brother, the beautiful day intoxicates Marion: "All around was so beautiful and so pleasing that Marion was insensibly led on and on."[93] After two hours of walking through the Edenic landscape, Marion falls asleep under an oak beside a brook. In shades of Goodrich's story "The Faithful Dog," the family spaniel, Hart, tracks her and finding her asleep, stays by her side guarding her as she sleeps.

Soon after Marion falls asleep, a "catamount" (panther) emerges from the forest to attack the little girl. Largely reflecting Cooper's panther attack scene in *The Pioneers*, the family dog battles the panther, delaying its attack on Marion, and in just the nick of time her brother appears from the woods to shoot the predator, just as Natty Bumppo does in *The Pioneers*. What is more interesting than the panther attack itself is Goodrich's portrayal of the wild animal's reputation and disposition. He writes,

---

90 Goodrich, *Peter Parley's Story of Little Marion*, 4–5.

91 Goodrich, *Peter Parley's Story of Little Marion*, 5.

92 Goodrich, *Peter Parley's Story of Little Marion*, 6.

93 Goodrich, *Peter Parley's Story of Little Marion*, 7.

> The catamount, I suppose you know, is a fierce and cruel animal, shaped like a cat and as large as a dog. It is a beautiful animal, of a fawn color, with black spots or stripes upon it. ... The animal is cowardly and will fly from a man; but it kills hares, rabbits and other small animals, and if it meets a child, it does not fear to attack it.[94]

Hence, while Goodrich presents his young readership with an advocacy of exploring the natural world and appreciating its beauty, he warns them of the dangers therein by furthering a conception of top predators as not simply dangerous but of poor character, even evil. As was the norm in the early nineteenth century, and as is often the case today, the predator is anthropomorphized not as a rightful citizen of the woods but as a criminal threat, a figure that must be exterminated if humans are to enjoy the mankind-fashioned pastoral that figures as the rightful natural world. Driving home this point, Goodrich, via the narrating Parley, says, "Let us all be thankful, my dear children, that we live in a land where there are no wild beasts."[95]

Sentiments such as these repeatedly emerge in Goodrich's work and reveal him, despite his sometimes frustrating ambivalence, as an important proto-environmentalist writer. He offers beautiful descriptions of the natural world replete with sentimental glances back to the lush forests of his childhood, and he cautions children to treat certain animals (but not all) with kindness. He links the grandeur of the North American natural world with the promise of the still fledgling United States, even if that promise is in his mind always linked to its potential for commodification. Still, we should be careful not to condemn Goodrich for simply adhering to the environmental views so prevalent both in his day and still in play in the twenty-first century. A reading of his juvenile literature reveals a writer who succeeded in combining his love of country, children, and the natural world into a profitable and educational series of books.

Conclusion

Early American environmental juvenile literature (specifically from 1780 to 1833) promoted a duplicitous human relationship with the natural world. On the one hand the interest in natural histories endorsed an early appreciation of facts about the lives and ecological well-being of animals. But those same natural histories—like much of the juvenile literature of that time—were almost always couched in Christian rhetoric, arguing that animals were proof of God's design, and made, along with other natural elements, for human purposes. The 1820s and 1830s saw a significant increase in books by and for Americans, and many of these texts were full of beautiful descriptions of the natural world. Nevertheless these books also present the idea that the environment, while beautiful, must—like the American Indians—fall to a process of largely unquestioned expansion of European American

---

94 Goodrich, *Peter Parley's Story of Little Marion*, 9–10.

95 Goodrich, *Peter Parley's Story of Little Marion*, 14.

interests, which run the gamut of environmental degradation—reshaping the natural world and eradicating species that might prove inconvenient to human interests (such as panthers and wolves) or whose commercial value is high (for example, passenger pigeons and bison). Samuel Griswold Goodrich, because of his intense popularity, had a significant impact on the environmental imaginations of American children, and despite the predictable environmental contradictions found in his work, he still deserves credit for writing juvenile books that overtly and sympathetically presented the American natural world to the nation's children. Goodrich's young readers would take his lessons with them as they became adults and went on to foster the ascension of the United States to cultural and industrial prominence in the second half of the nineteenth century.

PART II
Wild Visions

One day America will be as beautiful in actuality as it is in Cooper. Not yet, however. When the factories have fallen down again. (56)

—D. H. Lawrence, Studies in Classic American Literature (1923)

Chapter 4
Speculation, Degradation, and *The Pioneers*

> this speculation lies in the province of fancy and may be treated as a vision.
>
> —William Cooper, on his idea for the creation of an Erie canal, *A Guide in the Wilderness* (1810)[1]

Ralph Waldo Emerson's masterpiece of environmental philosophy, *Nature*, would not emerge until 1836. Henry David Thoreau's *Walden* would not appear until 1854, and George Perkins Marsh's *Man and Nature* would not emerge until 10 years after that. But years before them all, in 1823, James Fenimore Cooper published *The Pioneers*—a novel of remarkable environmental foresight. Reportedly selling 3,500 copies on its first day of release,[2] *The Pioneers* was, from the outset, an immensely popular book in both the United States and Europe. It has never since been out of print. This popularity no doubt made *The Pioneers* one of the most read of early environmental texts, one that is still read and taught today, and one that unfortunately becomes increasingly relevant as humanity continues to abuse the thin glaze of life that encases the Earth.

If, after the previous chapter, we need more proof of how formative the youthful reading of some of our most prominent nineteenth-century thinkers could be, we need look no further than the brief note Emerson sent to the 1852 memorial service that followed Fenimore Cooper's death. Emerson wrote, "I have, in common with almost all who speak English, an old debt to him of happy days, on the first appearance of the Pioneers."[3] While the book's actual influence on Emerson's writing is unknown, it is pleasant to think that it had at least some impact on the then 20-year-old's thinking. Herman Melville, who just the year before had published *Moby-Dick*, sent a letter as well, and spoke to Cooper's deep influence: "[H]is works are among the earliest I can remember, as in my boyhood producing a vivid and awakening power upon my mind."[4] It was, after all, Cooper, who with *The Pilot* (1821) had invented the maritime novel.

Just as Cooper's *The Pioneers* influenced these authors during their formative years, so too did it (and his other wilderness romances) impact the trajectory of the national environmental mind, even if that impact has been at times more latent than manifest. The purpose of this chapter, however, is not to argue for *The Pioneers* as an environmental novel; anyone who actually reads the book will quickly come to

---

1 Cooper, *A Guide in the Wilderness*, 23. Subsequent references appear parenthetically in the text.

2 Dekker and McWilliams, "Introduction," *Fenimore Cooper: The Critical Heritage*, 1.

3 *Memorial of James Fenimore Cooper*, 33.

4 *Memorial of James Fenimore Cooper*, 30.

that conclusion.[5] Instead, I aim to examine how Cooper's land-speculator father, William, contributed to the book's environmental stance, and to illuminate how *The Pioneers* serves as a criticism of the kind of land speculation philosophy that William—and indeed the nation—so disastrously embraced.[6]

William Cooper's Speculative Vision

The Pioneers bears obvious corollaries between the book's Temple clan and Cooper's family history, with Cooper forging Marmaduke Temple in the image of his own legendary father, William. The frontier village of Templeton is the mirror image of Cooperstown, New York, reflected on the surface of Lake Otsego, a body of water found at the foot of both the fictional and real towns. Given these associations, it is tempting to class *The Pioneers* as merely a nostalgic reimagining of the Cooper family's past glory, a backward glance across the years to the peak of William Cooper's improbable ascendance from impoverished wheelwright to wealthy land speculator. That corollary, however, would be an oversimplification, one that Alan Taylor makes when he writes that Cooper's *The Pioneers* is "the culmination of the Coopers' efforts to achieve wealth and authority."[7] Claiming the novel functions as a final gesture in the Cooper family's pretentions to glory, Taylor argues that in the book Cooper aims "to reassert his family's claims to preeminence."[8] Such a view fails, however, to appreciate the complexity of the younger Cooper's reimagining of his father and of the significance of frontier Cooperstown as a microcosm of the American colonial imperative. This take on the novel particularly underserves the role of Marmaduke Temple, who Cooper presents as the embodiment of a well-intentioned but often contradictory American land-use philosophy. That same philosophy can be found both in William Cooper's life and in his 1810 treatise on land speculation, *A Guide in the Wilderness*. Despite the initial success of William Cooper's land speculation, he ultimately

---

5 The critical recognition of *The Pioneers* as an environmental text has a long history. See, for example, Robinson's 1967 *PMLA* article, "Conservation in Cooper's *The Pioneers*." The recent *Encyclopedia of the Environment in American Literature* includes a pair of substantial entries on Fenimore Cooper, one biographical and the other on his Leather-Stocking Tales, and each stress the value of his novels as seminal environmental works; see Hamilton and Jones, 68–70.

6 Interestingly enough, *The Last of the Mohicans* seems to have garnered as much, if not more, attention from ecocritics, despite that novel's less overt environmentalism. For example, David Mazel, in his instructive study of Cooper's use of the environment in *The Last of the Mohicans*, contends that the book "repeatedly assigns specific and ideologically responsive meanings to wild nature—that is, it writes the wilderness ... the novel is spectacularly successful, prodigiously inventive in its encoding of ideology into nature" (72), an idea that might be assigned to *The Pioneers* as well.

7 Taylor, *William Cooper's Town*, 8.

8 Taylor, *William Cooper's Town*, 8.

failed to provide either familial or community stability, and his son's novel locates a significant portion of that failure in a loss of environmental integrity.

The Pioneers, then, is a lesson in the failure of America's speculative vision, for all of its economic genius, to account for long-term environmental consequences. It is a trait found at the core of Judge Temple, who repeatedly displays a genuine desire to preserve the beauty and integrity of the land while at the same time envisioning a future of radical change in the form of human development. As Cooper writes,

> The mind of Judge Temple, at all times comprehensive, had received, from his peculiar occupations, a bias to look far into futurity, in his speculations on the improvements that posterity were to make in his lands. To his eye, where others saw nothing but a wilderness, towns, manufactories, bridges, canals, mines, and all the other resources of an old country were constantly presenting themselves, though his good sense suppressed, in some degree, the exhibition of these expectations.[9]

Thus Marmaduke Temple figures not only for William Cooper but also for the new nation's failure to reconcile two powerful environmental urges. Looking at the land, European Americans saw a realm that they associated, in its wildness and beauty, with their idealized selves, and that formed the myth of a nation of rugged, uncompromising people. Yet in that same land they also saw the power of unused resources, and the vision of a new nation built upon a conquered wilderness. By addressing this conflict, one that may be found both in the narrative itself and at the heart of figures such as Marmaduke Temple and Natty Bumppo, *The Pioneers* became not only one of the first but also one of the most important works of nineteenth-century environmental fiction. As Timothy Sweet writes,

> On the one hand, Cooper imagines the possibility of an escape from economy altogether in the figure of Natty, who ... seems to embody a nonexploitative relationship to the natural environment. On the other hand, within the terms of the settlement economy [represented by Temple], Cooper addresses issues of law, property, and practical environmental management.[10]

Fenimore Cooper's prescient engagement with the problem of environmental exploitation derives in part from his father's ideas about successful land speculation. As detailed in William Cooper's posthumously published, *A Guide in the Wilderness*, one key to his success had been his mastery of environmental prophesy. Levying his speculative gaze upon the New York wilderness, he predicted vital trading towns would emerge at "the mouth of Buffalo Creek" and at "the first falls of the Genesee" (21), foreseeing, respectively, the future settlements of Buffalo and Rochester. His eye for economic development depended upon

---

9 Cooper, *The Pioneers*, 321. Subsequent references appear parenthetically in the text.

10 Sweet, *American Georgics*, 154.

reading the land's potential for human exploitation. Drawing upon his father's land-use philosophy, Cooper infused the town of Templeton with characters and events that exemplify the need to modify the nation's speculative vision so as to recognize how environmental loss follows shortsighted exploitation. By its end, however, the novel concedes that, when pitted against the capitalistic urge, such preservationist warnings do little to stem environmental degradation. The potential for individual gain becomes simply too great to ignore, and few figures serve as a better embodiment of this mindset than William Cooper himself.

The elder Cooper was nothing if not entrepreneurial. Driven by an urge to overcome his modest origins as the son of an impoverished farmer, he left home to become a wheelwright.[11] At the age of 20 he married Elizabeth Fenimore, the daughter of a wealthy farmer. Her wealth allowed William to begin to enter the ranks of the upper class, but he knew he also needed social refinement, education, and connections with members of the community's elite citizenry. He found the gateway to each of these necessities through his membership with the Library Company of Burlington, which not only gave him access to approximately 1,400 volumes—319 of which he read over the next 10 years—but to the Library's highly educated and well-heeled members.[12] Combining this vigorous self-education and social networking with a talent for land speculation, he began to rise in the social ranks, and with this rise in class came further opportunities, the greatest of which occurred in about 1785. That year—through connections from his membership with the Library Company—he learned that a group of investors were looking to sell the so-called Otsego Patent, a 29,350 acre tract that spread west from Lake Otsego into the frontier of central New York. As Alan Taylor explains, "such a substantial tract in the path of the westward migration of New Englanders could enrich whoever obtained title and retailed farms to settlers."[13]

Eager to see for himself the promise of this land that lay so profitably in the path of westward expansion, William, on horseback, rode into the wilderness of central New York. In *A Guide in the Wilderness*, he recalls first seeing the land that would soon after become his to control. Clad in an almost prophetic tone, his description of the land as he found it reveals both the complexity and the blind spots of his speculative gaze. He writes,

> In 1785 I visited the rough and hilly country of Otsego, where there existed not an inhabitant, nor any trace of a road; I was alone, three hundred miles from home, without bread, meat, or food of any kind; fire and fishing tackle were my only means of subsistence ... I laid me down to sleep in my watch coat, nothing but the melancholy Wilderness around me. In this way I explored the country, formed my plans of future settlement, and mediated upon the spot where a place of trade or village should afterwards be established. (9)

---

11 Taylor, *William Cooper's Town*, 16.

12 Taylor, *William Cooper's Town*, 22–3.

13 Taylor, *William Cooper's Town*, 28.

He soon found the place he was hunting for. Scaling a mountain overlooking Lake Otsego, he gazed across the land and saw from his high elevation the future site of Cooperstown, the canvas upon which he would enact his speculative will. As adept as he was in imagining the addition of the intrusive trappings of settlement, he was also a master of omission. He claims that in the Otsego country of 1785 "there existed not an inhabitant," but the area was at various times inhabited both by other European Americans and, most clearly it was still at that time well known to the Iroquois, whose ruins from past settlements dotted the landscape. A mere 20 years before William Cooper wrapped himself in his watch coat and bedded down for the night among the "melancholy Wilderness," a group of Christianized Mohawks had tried to make a permanent settlement on the site where Cooperstown itself now stands. Even after they abandoned that plan, Indians still frequented the area to hunt and fish.[14] "In recounting his vision," writes Taylor, "William Cooper erased all of the land's previous owners and users ... He implied that no predecessor, Indian or white, had owned or worked the land."[15] In a gesture fitting his prophetic self-perception, William named his lofty vantage point Mount Vision, and, as Taylor observes, "Thereafter Mount Vision assumed a special importance in the Cooper family memory."[16]

William's omission of the land's former, and, in some cases present occupants, reveals the extent to which his speculative gaze was a product and promoter of the larger national vision that depended too often upon the removal of inconvenient inhabitants, and often in the displacement of entire communities. William's vision rejects the land's human past and mirrors the spirit of removal found even in a mere passing glance from the colonial speculative gaze. With this figurative erasure of the former inhabitants, as well as the omission of the traces of their past occupancy, William reinvokes the well-worn fiction of the virgin wilderness. One of the most persistent of American myths, the idea of the unspoiled wilderness far predates William Cooper, but in his descriptions of the Otsego Patent we hear the echo of this national delusion of an untouched wilderness that served as both pleasure ground and proving ground for the European American fantasy. "[W]ilderness," writes William Cronon,

> is not quite what it seems. Far from being the one place on earth that stands apart from humanity, it is quite profoundly a human creation ... It is not a pristine sanctuary where the last remnant of an untouched, endangered, but still transcendent nature can for at least a little while longer be encountered without the contaminating taint of civilization. Instead, it is the product of that civilization.[17]

14 Taylor, *William Cooper's Town*, 36–7.

15 Taylor, *William Cooper's Town*, 34.

16 Taylor, *William Cooper's Town*, 32.

17 Cronon, "The Trouble with Wilderness," 69.

And one of the most effective promoters of such a socially constructed wilderness was William's son, whose Leather-Stocking Tales and other romances so fully incorporated it into the American myth.

Decades after William Cooper's account, James Fenimore Cooper reimagined his father's first glimpse of the area. In *The Pioneers*, Marmaduke Temple looks back upon his discovery of the site for Templeton, and his description bears close consideration, both for its debt to William Cooper's relation of the event in *A Guide in the Wilderness* and for how it veers from that version. Unlike William Cooper, Marmaduke Temple opens by recognizing that he is not so entirely alone in the wilderness as he may seem but is instead part of a relatively nearby group. He says,

> I left my party, the morning of my arrival, near the farms of the Cherry Valley, and, following a deer-path, rode to the summit of the mountain, that I have since called Mount Vision; for the sight that there met my eyes seemed to me as the deceptions of a dream. (235)

He then relates, in great and dramatic detail, a landscape far richer and far more complex in character than William's portrayal:

> The fire had run over the pinnacle, and, in a great measure, laid open the view. The leaves were fallen, and I mounted a tree, and sat for an hour looking on the silent wilderness. Not an opening was to be seen in the boundless forest, except where the lake lay, like a mirror of glass. The water was covered by myriads of the wild-fowl that migrate with the changes of the season; and, while in my situation on the branch of the beech, I saw a bear, with her cubs, descend to the shore to drink. (235)

Cooper writes of Marmaduke Temple's view from Mount Vision in language that at first may seem essentially identical to that of his father's account, and according to Taylor, the younger Cooper "recurrently employed the view from Mount Vision as a symbolic vantage point for assessing the troubled evolution of the community below" (32). Taylor is certainly correct that James redeploys this anecdote as commentary, but his repurposing of the scene has more to do with the way that William Cooper and Judge Temple saw and thus figuratively possessed the land than with the changes in the human community itself. Cooper employs this kind of nuanced, editorial description throughout his Leather-Stocking Tales, but it is often overshadowed by the mere power of his descriptions. For example, in his study of Cooper's use of the environment as a masculine and nationalistic proving ground in *The Last of the Mohicans*, David Mazel writes, "[T]he literary environmentalism of *The Last of the Mohicans* depends less upon Cooper's love of nature than it does upon his novelistic technique. And the crucial aspect of that technique ... is his handling of the convention of the picturesque."[18] This assertion certainly rings true for *The Last of the Mohicans* as well as for *The Pathfinder*, and

---

[18] Mazel, *American Literary Environmentalism*, 65.

The Deerslayer, but Cooper's approach to the picturesque in *The Pioneers* and *The Prairie* is more conscientiously environmental. It is more about reading between the lines of the landscape than merely painting a pretty picture.

In *The Pioneers*, William Cooper's vision becomes James Fenimore Cooper's *revision*, and the differences between the two accounts figure as the son's commentary on the weaknesses inherent in his father's lofty speculative gaze. While both passages evoke a majestic and lonely wilderness, where William Cooper saw mostly absences, Marmaduke Temple sees a land filled with animals, and, perhaps even more significantly, he recognizes their movements through the seasons. Temple's wilderness is alive and changing, and in that respect it mirrors the narrative of *The Pioneers* itself, in that it is a drama that follows the stage direction of the seasons and tells the story not only of bears and deer and water-fowl but also of a diversity of remarkable people. One of those people is Natty Bumppo, whose cabin he chances across after he descends from his lofty view on Mount Vision.

That Temple only sees Natty, or any human inhabitant, until after he climbs down from the mountain is one of a few hints that Cooper is not only revising but also criticizing this family legend. While on the mountain, Temple is blind to the human inhabitation of the land. He admits,

> not the vestige of a man could I trace, during my progress, nor from my elevated observatory. No clearing, no hut, none of the winding roads that are now to be seen ... nothing but mountains rising behind mountains ... Even the Susquehanna was then hid, by the height and density of the forest. (235–6)

In other words, while this elevated, privileged view of the land is beautiful, it results in a form of impaired vision because he is so far removed from the intricacies and workings of the land as to be oblivious even to the Susquehanna River meandering through the forest below. William, however, in his description from *A Guide in the Wilderness*, never accepts the human history and presence on the land, even when it was very clearly before him. If, as Taylor argues, William erases any previous human claims to the land, then his son, in *The Pioneers*, is engaged in "restoring the Indians, squatters, and colonial landlords," and this gesture, as Taylor suggests, "imparts a very different cast to William Cooper's story, shifting his ownership and development of Otsego from the beginning to the midst of a history of conquest and settlement."[19] This statement seems to contradict Taylor's previous claim that *The Pioneers* represents "the culmination of the Coopers' efforts to achieve wealth and authority,"[20] and instead contends that the novel is as much a recasting of William Cooper as it is a reimagining of the land that so fired his entrepreneurial mind. Even as Cooper repopulates Templeton with the Indians and settlers that his father omitted from his account, it also re-envisions William Cooper himself, giving him characteristics perhaps underappreciated or even absent in actuality.

---

19 Taylor, *William Cooper's Town*, 34.

20 Taylor, *William Cooper's Town*, 8.

And, as we will see, perhaps the most intriguing character trait *The Pioneers* gives to Marmaduke Temple, in its oblique reformation of William Cooper, is a decidedly environmental consciousness coupled almost ironically with a visionary genius for land speculation.

When William looks out across the land of the Otsego Patent, he sees a paradise of potential economic development, of villages planted at the feet of lakes, of roads born of deer paths, of cities moored on riverbanks. It is, after all, no surprise that an early republic land speculator would demonstrate much less concern for the integrity and health of the land than for its potential for human gain, especially his own. For a man like William Cooper, the ideal landscape was not a wilderness or a well-tended garden but a well-oiled commercial enterprise that leveraged substantial tracts of land to produce goods marketed to the world beyond. The goal then for William was to carve this commercial settlement of productive farms and well-governed villages, of shepherds and roadways, from a formerly perilous land. He then fits what Roderick Nash terms "the pioneer's association of wilderness with hardship and danger in a variety of forms," although in William Cooper's case perhaps the more appropriate designation would be that of "developer" instead of "pioneer."[21] (In fact, Cooper biographer, Wayne Franklin, notes that while Cooper's selection of book title now seems exceedingly commonplace, it was anything but common at the time.)[22] "[T]he rural, controlled, state of nature," writes Nash, "was the object of his [the pioneer's] affection and goal of his labor. The pastoral condition seemed closest to paradise and the life of ease and contentment. Americans hardly needed reminding that Eden had been a garden ... On both the idyllic and practical counts wilderness was anathema."[23] As true as this statement is, what sets William Cooper apart is that while he did profit from the development of the New York wilderness, he was not without a certain amount of regret for the environmental degradation that followed. Referring specifically to William Cooper's settlement of the Otsego Patent as an example of this mindset, Nash points to the opening pages of *A Guide in the Wilderness*, where William writes, that all parties involved in the settlement of land share the

---

[21] Nash, *Wilderness and the American Mind*, 30.

[22] As Wayne Franklin explains, Cooper's referring to his characters as pioneers had more to do with a definition that has since been eclipsed by the very popularity of the book itself, a case of an author's success redefining the very terms employed in creating it. Franklin writes that when Cooper described his characters

> as "our precursors or pioneers," he hardly was using the latter word in the heroic guide is later would acquire. Instead, fully aware of the military origin of the term (in French, les pionniers were an army's advance guard, its shock troops), he was emphasizing the role the first settlers played in what he also called the "march of the Europeans toward the interior parts of the continent."

The ensuing popularity of Cooper's book would forever transform the term from its relatively obscure origins into what we now consider a catchall designator for settlers of rugged terrain; see Franklin, *James Fenimore Cooper: The Early Years*, 338.

[23] Nash, *Wilderness and the American Mind*, 30–31.

same "great primary object ... to cause the Wilderness to bloom and fructify" (7). But William also, in rare but notable moments, displays the glimmer of a conservationist ethic. Just a few paragraphs later he writes that the typical settler of a parcel of land "clears away the trees until they, which were the first obstacles to his improvement, becoming scarcer, become more valuable, and he is at length as anxious to preserve as he was at first to destroy them" (7). William's example displays the preservationist urge as the product of scarcity itself, as the inevitable and well-known economic consequence of the depletion of a resource, but what is notable about his account is the manner in which the recognition of that depletion betrays the modest beginnings of a larger environmental consciousness, one that has as its concern not only the short-term sale and development of land but also the long-term consequences of that process.

William Cooper's comments about the near inevitability of deforestation represent a larger cultural recognition of how the land was changing due the impact of European American development, a recognition that became more common as the nation made its way into the nineteenth century. In her study of the arc of American environmental history, Carolyn Merchant writes that in the earliest years of settlement, wilderness had been viewed as an obstacle, as an abode of evil that served only to fuel the larger colonial project through its taming and the use of what seemed at the time an inexhaustible abundance of natural resources. This wasteful and antagonistic view was the product of the Puritan mindset, by the ignorance of those earliest European settlers who saw everywhere in the new landscape proof of God's presence on earth, and who derived their land-use ethic from a biblical source that had nothing to do with the ecology and nature of North America. Merchant writes,

> the concept of wilderness came with the Puritans to the New World as cultural baggage. The wilderness is the antithesis of the garden, or of paradise. The biblical story says that Adam and Eve were driven out of the Garden of Eden into the wilderness, a desert. They had to labor in the earth to change it into a garden, just as the colonists had to change the New England wilderness into a cultivated garden.[24]

As evidenced by the writings of those such as William Cooper, this perception had, over two centuries of cultural and environmental change, shifted to a more positive light as the human development of the land made dense forests and unspoiled landscapes far less common, and therefore increasing both the economic and aesthetic value of what were once merely obstacles to a second Eden. "The American wilderness," writes Merchant, "began to be appreciated in the nineteenth-century as the nation's forests began to disappear. What remained of nature that had not been used for economic purposes or settled was eulogized as wilderness."[25]

---

[24] Merchant, *American Environmental History*, 35.

[25] Merchant, *American Environmental History*, 36.

Still, it would take several decades before this growing cultural recognition of deforestation's ills emerged as an articulate, conservationist statement. While he was not the first to come to this realization, George Perkins Marsh perhaps conveyed it best when he promoted the idea that maintaining tree cover was essential to both community and environmental stability. Echoes of William and James Cooper are heard in Marsh's *Man and Nature* (1864) when he writes:

> [W]hen we consider the immense collateral advantages … the preservation of existing woods, and the far more costly extension of them where they have been unduly reduced, are among the most obvious of the duties which this age owes to those that are to come after it.[26]

But this idea, so well put by Marsh and so well revised and reinvigorated by Aldo Leopold in the following century has origins far older than the 1860s. This train of environmental thought—this idea that deforestation threatens the very ecological and agricultural integrity of the land—can be found in the earlier writings of forwarding-looking agriculturalists such as Jesse Buel, who, in 1839, wrote

> [I]n our zeal to *clear up*, we generally carry the matter to an unwarrantable extreme; every thing is cut away—the whole surface is denuded—stripped of its natural growth ... [trees] if left in clumps and belts upon ... portions of the farm not adapted to ploughland or to meadow, would tend ultimately to enhance its value, by the beauty which they would impart to the landscape, the shelter and protection which they would give to crops and cattle, and by the resources they would give for fuel, fencing, and timber. The settler upon new lands may *preserve*, without labor or expense, that which it would cost much time and money to *produce* ... To destroy, in this case, is but the labor of a day; to restore, is the work of an age.[27]

This sentiment is found still earlier in the writings of farmer John Lorain, who in his *Nature and Reason Harmonized in the Practice of Husbandry* (1825) also cautioned against a shortsighted view of land use when it came to the value of wooded land in association with farms:

> In new settlements, the timber is commonly the greatest obstacle to cultivation; this begets an emulation to destroy it; he is considered the best farmer, who clears the most land. This enterprise would be laudable to a certain extent; the habit, however, of considering the timber in the way, induces the farmer to wage a perpetual war against it, until his eyes are opened, by finding that he has neither fencing nor firewood left.[28]

---

[26] Marsh, *Man and Nature*, 327.

[27] Buel, *The Farmer's Companion*, 55.

[28] Lorain, *Nature and Reason Harmonized in the Practice of Husbandry*, 333.

Steven Stoll argues that when these agricultural writers are taken into account, Marsh's preservationist sentiments appear more a rearticulation of what agricultural thinkers had been saying for years.[29] Buel's and Lorain's warnings were, however, preempted by Cooper in 1823 and William Cooper in 1810, and indeed, while Lorain, Buel, and Marsh had probably never read *A Guide in the Wilderness*, it is tempting to consider that at least one of the three had read and taken to heart the environmental lessons of the internationally bestselling *The Pioneers*.

William Cooper's brief mention of how deforestation leads to a higher valuing of those trees that remain gives rise to an embryonic understanding of the results of environmental degradation. His son James then helps complete the process described by Merchant by writing a novel that eulogizes the lost wilderness of the Otsego Patent. In doing so, the younger Cooper cultivates and magnifies the spark of conservationist feeling found in his father's writings into the forward-thinking proto-environmentalism of *The Pioneers*, a book that takes as one of its main concerns the problem of how human settlement and exploitation of the land leads to irreversible environmental loss.

The Environmental Prescience of *The Pioneers*

Despite *The Pioneers*'s status as a major bestseller with an overt conservationist message, there remains a perplexing resistance by some scholars to accept its importance as a key early environmental text. A relatively recent example of this critical tendency is found in Bradley P. Dean's article, "Natural History, Romanticism, and Thoreau," in which he writes,

> In 1823 ... Cooper published *The Pioneers*, the first of his enormously popular Leatherstocking Tales and the book that established him as the country's first bestselling novelist. Despite the impressive achievements of Cole and Cooper, however, America's cultural independence was not achieved until shortly after ... Emerson moved to Concord in 1834. His seminal book *Nature* (1836), widely regarded as the first literary and philosophically sophisticated articulation of distinctively American ideas, inaugurated what has since become known as the American Renaissance.[30]

I will set aside the problematic claims that European America could ever obtain "cultural independence," from its European origins; that Emerson's ideas—heavily influenced by the German Philosopher Immanuel Kant—were somehow "distinctively American;" and that Cooper was "the country's first bestselling novelist."[31] The far more troubling issue with Dean's conception of American

[29] Stoll, "*Farm against Forest*," 66.

[30] Dean, "Natural History, Romanticism, and Thoreau," 76.

[31] America's first best-selling novelist, itself mostly a subjective honor, belongs to Susanna Rowson, whose 1791 sentimental novel, *Charlotte Temple*, resounded with American readers. See Davidson, *Revolution and the Word*, 42.

literature (one that perhaps speaks to the larger critical problem of ignoring the impact of Cooper's work), is the tendency of some critics to adhere too strongly to F.O. Matheson's outdated concept of the so-called American Renaissance. This idea, with its origins in mid-twentieth-century literary thought, suggests that despite the influence of so many pre-1836 works by authors such as Charles Brockden Brown, J. Hector St. John de Crèvecoeur, Cooper, Washington Irving, Catherine Maria Sedgwick, and many others, that American literature somehow really only began after the publication of Emerson's *Nature*. Predictably, these critics tend to mention Cooper's work only in passing before moving on to Emerson, and then to Thoreau's *Walden*. While those later masters certainly deserve the scholarly attention they have received, the critical underserving of Cooper's work is an odd tendency, especially given his portrayal in *The Pioneers* of the human impact on the natural world, its conservationist-minded characters of Natty Bumppo and Marmaduke Temple, and its overarching and sophisticated environmentalist message. Other than an outdated conception of American literary history, perhaps this critical oversight results from a lack of knowledge of the text itself (yet even a cursory reading reveals the novel's strong environmental slant), or perhaps it is the product of a tendency to consider the book as low culture when placed in comparison with Emerson's later text. An excuse based on concepts of low versus high culture would, however, actually argue for *The Pioneers* as the book with the far wider social impact, than say, Emerson's less accessible and probably far less read *Nature*.

Likewise, Aaron Sachs's recent study of the pastoral in nineteenth-century American culture strangely downplays James Fenimore Cooper's role in the evolution of American environmental thinking. He correctly lauds Washington Irving and Susan Fenimore Cooper as writers who helped put the landscapes of a romanticized New York on the cultural map, but then he writes, "James Fenimore Cooper also had a great deal to do with creating the iconic status of upstate New York, but ultimately he was less influential than Washington Irving and less of an Arcadian than his daughter Susan."[32] Irving's *Sketchbook of Geoffrey Crayon* was indeed highly influential, but no more so than any one of Cooper's far more developed bestsellers of the 1820s, especially *The Spy* (1821), *The Pioneers*, and *The Last of the Mohicans*, all of which incorporate a compellingly imagined New York landscape. And Susan Fenimore Cooper's *Rural Hours* is definitely Arcadian, but no more so than *The Pioneers* and, indeed, can be read as a companion volume to her father's novel.

Happily, the environmental facets of Cooper's work itself have not gone unnoticed by several other recent critics. Bryan L. Moore contends that "Cooper's novels, taken as a whole, reflect a disillusionment with the development of the wilds, and his major novels indicate a shift in the perception of America's plentitude."[33] Michael Ziser, amplifies and expands upon this idea in his study

---

32 Sachs, *Arcadian America*, 409 n.3.

33 Bryan L. Moore, *Ecology and Literature*, 110–111.

of the portrayal of environmental practice (specifically bee hunting) in Cooper's *Oak Openings*. In a sentiment that easily applies to virtually all of Cooper's forest romances, Ziser argues,

> *Oak Openings*' main work is to show how the path from the material to the symbolic is laboriously traveled through the everyday exercise of environmental practice. Cooper was equipped, as few are, to appreciate the reciprocal trespass of nature and culture—bees and trains—on one another and on the properties of his mind.[34]

And in one of the strongest analyses of Cooper's environmental literary agenda, Lloyd Willis notes how a presentist critical condemnation of Cooper's work (in large part as a result of F.O. Matthiessen's and George Santayana's work) undeservedly "devalued Cooper, which has had the secondary effect of silencing the environmental anxieties that are fundamental to his body of work even as it preserved the centrality of Emerson and his philosophy of environmental evasion."[35]

Much of the criticism that does address Cooper's take on environmental degradation in *The Pioneers* has focused on Natty Bumppo, especially those studies that explore the book's engagement with the ideas of nationalistic identity, wilderness, freedom, and social responsibility. It is easy to understand why. In Natty, Cooper invented (perhaps without fully realizing the character's power) a figure who embodies much of the American myth. Cooper soon realized he was on to something with Natty, and as the later four Leather-Stocking Tales attest, Cooper uses the variously named but consistently compelling character to plumb the American social consciousness. It is a telling fact that of all of the characters in *The Pioneers*, it is Natty who rose from the crowed to thrill the reading public. The irascible woodsman's nickname "Leather-Stocking" has since become shorthand for a powerful, if contradictory, American mindset.

In his history of trees and forests in American culture, Eric Rutkow points to Cooper's invention of Natty Bumppo as a significant moment in the development of an American environmental identity. Rutkow writes that Natty serves as "Cooper's new universal icon of America," and that ultimately,

> the sociohistorical elements of the backwoodsman (pioneer forest hunters; men of quasi-nomadic disposition; blazers of wilderness trails for the incoming waves of settlers) and the fictional-mythological qualities (limitless valor; unrivaled marksmanship; disdain for society; oneness with the forest and its trees) were competing to form a new American hero.[36]

---

34 Ziser, *Environmental Practice and Early American Literature*, 156.

35 Willis, *Environmental Evasion*, 38–9.

36 Rutkow, *American Canopy*, 67.

Notably absent from Rutkow's assessment of the character of Natty is any actual conservationist leaning. While this mythical frontier hero certainly voices such concerns in *The Pioneers* and *The Prairie*, Rutkow is correct in leaving out an environmental mindset when summing up the essence of Cooper's most famous character. It simply does not hold up as a character trait when we view Natty across the arc of the five Leather-stocking Tales. Nevertheless, it remains the Daniel Boone–inspired Natty who resonates so powerfully and permanently with the American public. As Henry Nash Smith writes, "the character of Leatherstocking is by far the most important symbol of the national experience of adventure across the continent."[37]

Natty is not, however, the novel's most important symbol of American environmental consciousness. After all, even in *The Pioneers*, he is better at interpreting the land than he is at acting as its savior. Though on the surface, he seems to strongly represent the need for a socially responsible relationship with the land (and he is certainly more sympathetic to the natural world than most of the characters who populate the book) there is a telling and even dangerous short-sightedness to his environmental philosophy. He simply cannot realistically address the complex problems of developing a sustainable relationship between a human community and the land. When he sees the villagers overfishing Lake Otsego, Natty appears just to chide them and (showing off) demonstrates his method of spearing a single fish. When they shoot masses of passenger pigeons from the sky, Natty demonstrates that he can shoot a single bird as needed, even as thousands lay dying on the ground. And when Temple expresses his support of a newly enacted law "prohibiting the drawing of seines, at any other than proper seasons, in certain of our streams and small lakes; and another, to prohibit the killing of deer in the teeming months" (160), Natty exclaims his derision of what he views as unenforceable legislation, "You may make your laws, Judge ... but who will you find to watch the mountains through the long summer days, or the lakes at night? Game is game, and he who finds may kill; that has been the law in these mountains for forty years" (160). Temple counters that the law will still have a positive effect if properly administered and that game is the property of the landholder: "I hope to live to see the day, when a man's rights in his game shall be as much respected as his title to his farm" (160). Natty, who abhors the idea of landowners holding sole ownership of their land and game, retorts, "[I]t's the farmers that makes the game scearce, and not the hunters" (161).

For Natty, settlement itself is the problem, but it is a problem to which he offers no real solution other than to continue the unregulated use (and hence occupation) of the land by only expert woodsmen such as himself—a small group indeed. So in essence, Natty argues for a type of ownership of the land after all, the kind he claims "has been the law in these mountains for forty years" (161), that is, the kind that has been possible since the occupation of the land by European Americans. Natty's land-use solution is for everyone but him to head back east while he stays

[37] Henry Nash Smith, *Virgin Land: The American West as Symbol and Myth*, 61.

on to live as he wishes among those few Native Americans like John Mohegan whose claim to the land far exceeds a mere 40 years. Natty's alternatives speak more to his abilities as a hunter than to any practical way of allowing a community to live responsibly on the land, and by the end of the novel it becomes apparent that his actual solution to any environmental problem is to simply leave, to head ever westward to another as-yet-unspoiled locale and wait for the wagons to follow. In fleeing settlers, Natty speeds settlement. That he does so while offering no realistic way to compromise between human use and environmental integrity underscores the hollowness if his environmental cautions, which ultimately amount to the idea that people should go away, or else he will, and in the book's final pages he does just that.

So while Natty does give voice to the idea of not engaging in so-called "Wasty Ways" (248), it is instead Marmaduke Temple who acts as the most practical environmentally minded figure in the book, though, like Natty, he too is burdened by moments of environmental short-sightedness and outright hypocrisy. For its time, Temple's environmental consciousness was notably sophisticated, in that he views the landscape as a site for a long-term, mutualistic relationship between humanity and a diverse environment. For Temple the ideal American community is neither a wilderness nor a city, but an in-between state that fosters a diversity of landscapes and an economic system of frontier exploitation that ensures a consistent payoff in goods. It is an ideal that resembles something like the current push to develop an environmentally responsible but economically lucrative mode of natural resource use, a trend often simply labeled as sustainability. This view becomes apparent early in the novel when Cooper describes how Temple reacts to the sight of his namesake town upon arriving there with his daughter Elizabeth on Christmas Eve. As his sleigh emerges from the forest the passengers gain a grand view of Templeton in the valley below, and the Judge "contemplated, with philanthropic pleasure, the prospect of affluence and comfort, that was expanding around him; the result of his own enterprise, and, much of it, the fruits of his own industry" (46). At this point in the narrative Temple's "enterprise" has started to take root in the form of this rustic town at the edge of Lake Otsego, and, for Temple, it is ostensibly an endeavor that is as much "philanthropic" as it is a form of ensuring "affluence and comfort."

Yet for Temple there is a limit to which that affluence and comfort should be tolerated, particularly when it impacts the sustainability of the environment. This proto-environmental consciousness hints at a significant difference in the land-use philosophies of the fictional Temple and his real-world antecedent, William Cooper. While William, at least in *A Guide in the Wilderness*, displays few conservationist intentions, the same is far from true for Temple, who along with Natty Bumppo gives voice to the need for a responsible approach to resource use. In one key example, as Judge Temple and his family and friends prepare to dine at his home, he notices Richard has stocked the hearth fire with sugar-maple logs. Temple exclaims:

> How often have I forbidden the use of the sugar-maple, in my dwelling. The sight of that sap, as it exudes with the heat, is painful to me, Richard. Really, it behooves the owner of woods so extensive as mine, to be cautious what example he sets his people, who are already felling the forests, as if no end could be found to their treasures, nor any limits to their extent. If we go on in this way, twenty years hence, we shall want fuel. (105)

Marmaduke Temple's conservationist mindset derives from a set of environmental concerns, each of which are rooted in economic and community stability. In lamenting Richard's choice of firewood as one that pains him as he watches the "sap, as it exudes with the heat," Temple illustrates his belief that a resource, here sugar maple trees, be reserved for its most renewable and economically profitable use, such as in the making of syrup. Temple then warns that the profligate waste of trees in general may lead to a day when they "shall want fuel." These two proposed outcomes, each the result of poor resource management, reveal his environmental foresight, one he displays at key moments in the text. (Yet it bears noting that Temple, in appointing the selfish and moronic Richard as his executive assistant, represents a dangerous blind spot in terms of leadership.)

At its core Temple's conservationist urge is a product of his economic ambitions; after all, he is speaking of his property and he demands that his sugar maple trees be reserved not for any ecological purpose but for their economic value. When Temple speaks of the issue of firewood, however, he is engaging in a separate, more community-oriented matter, and one that is no less impressive for its time. Warning the villagers that unless they engage in a conscientious use of natural resources they may see a day without firewood, Temple presents the kind of forethought becoming a leader of a precarious settlement. If he is to see his Templeton survive and prosper, he knows the people must make judicious use of its resources, even if those resources appear inexhaustible. Hence proper resource use helps ensure the success of the human community, even if concern for the non-human realm is largely ignored or subsumed within human interests. In short, Temple's environmental philosophy treasures the land and its natural features for their economic rather than aesthetic value, a view that is unsurprising and well within the speculative vision of someone like William Cooper. What is remarkable about Temple's environmental vision, however, is that he is painfully aware of the limits of these natural resources. Temple's environmental prescience relies upon a recasting of the same farsighted talents necessary for any successful land speculator: he looks across the landscape and envisions a possible future made bleak by poor land management and an ever-increasing human population. It must be stated, however, that for all of his virtues as a proto-environmentalist, Temple—in moments similar to those involving Natty—repeatedly engages in behaviors that overtly contradict good ecological sense. For example, he participates in Richard's wasteful netting of fish from the lake, and, like those he labels the "brats in the village," Natty too contributes to the slaughter of the passenger pigeons (246). That Temple feels guilty afterward does nothing to undo these environmentally irresponsible actions nor to repair the poor example the town's leader has set.

Another character, the woodchopper Billy Kirby, best embodies the problem of such a short-sighted, exploitative view of land use. In a passage recalling his earlier admonishment of Richard for heating his home with sugar maple logs, Temple and company come across Billy's camp. Despite his skill at various outdoor jobs, Billy, a Herculean woodsman, is most renowned for his prowess with an axe and claims to have cleared 500 acres all by himself. They find Billy busy tending two boiling kettles of maple sugar. Temple looks around disapprovingly and says,

> It grieves me to witness the extravagance that pervades this country ... where the settlers trifle with the blessings they might enjoy, with the prodigality of successful adventurers. You are not exempt from the censure yourself, Kirby, for you make dreadful wounds in these trees, where a small incision would effect the same object. I earnestly beg you will remember, that they are the growth of centuries, and when once gone, none living will see their loss remedied. (228)

In this day of sustainability, green initiatives, and Leadership in Energy and Environmental Design (LEED) certified campus buildings it is perhaps easy to overlook the far-sighted significance of Marmaduke Temple's warnings. In 1823 such conservationist ideas—whether applied to trees or any number of other natural resources—were exceedingly rare. In the 1820s most people probably agreed with the likes of Billy Kirby who in his response to Temple's admonition replies, "Why, I don't know, Judge ... It seems to me, if there's a plenty of any thing in this mountainous country, it's the trees. If there's any sin in chopping them, I've a pretty heavy account to settle" (229). Before leaving Billy's camp, Temple says, "it is not as ornaments that I value the noble trees of this country; it is for their usefulness. We are stripping the forests, as if a single year would replace what we destroy. But the hour approaches, when the laws will take notice of not only the woods, but the game they contain also" (229). Cooper forwards through Temple a savvy conservationist philosophy, one notable merely for its existence at such an early date, and one that has at least part of its origin in William Cooper's land speculation mindset. Still, Temple's moments of environmental waste and failed leadership must also be factored into Cooper's portrayal of the man meant as a stand-in for his father. When accounting for these weaknesses we realize just how critical Cooper is of his own family's environmental legacy, and just how wrong Allan Taylor is when he asserts that *The Pioneers* is "the culmination of the Coopers' efforts to achieve wealth and authority."[38]

However, not all critics would agree with this claim. Donald Ringe, for example, writes that Cooper portrays in *The Pioneers* the vicious but inevitable result of Judge Temple's sale of the land in that

> he initiates a process that takes on a life of its own ... White man supplants Indian, and agriculture displaces hunting as a means of livelihood. Though the

---

38 Taylor, *William Cooper's Town*, 8.

> cost may be regrettable, Cooper does not deny the legitimacy of the Judge's aims nor the value of the result he is trying to achieve.[39]

Ringe is right that Cooper does not himself overtly condemn the environmental degradation that followed settlement, but he did, however, create characters who do that job for him. Natty Bumppo repeatedly condemns the environmental damage and waste he sees, and he is given voice with such force and frequency that it is hard not to interpret his words as Cooper's own. Despite Ringe's claim, Judge Temple also repeatedly admonishes those around him who engage in wasteful resource management, a trait that makes him seem far from comfortable with the situation. He strongly condemns the wasteful and greedy practices of Richard Jones and Billy Kirby, and it is through these profligate figures no less than through the unapologetically conservationist Natty Bumppo that Cooper presents his repeated lament at the great human and environmental cost of poor land use. He even sheds these proxies in a note to the 1832 edition of *The Pioneers*, in which he writes: "Though forests still crown the mountains of Otsego, the bear, the wolf, and the panther are nearly strangers to them. Even the innocent deer is rarely seen bounding beneath their arches, for the rifle and the activity of the settlers have driven them to other haunts" (9). In reimagining the wildness of his childhood home, Cooper also reimagined the seeds of its destruction.

In *A Guide in the Wilderness*, William Cooper makes a handful of comments that gesture toward Judge Temple's forward-thinking (if still capitalistic) conservationist ethic. This view probably has at least some of its origins in the ideas of William, who had a talent, not only for predicting the best sites for the planting of towns, but also for anticipating how those settlements tended to change the land and with it the availability of resources. While William's environmental consciousness is not as sophisticated as that of his fictional counterpart, he was well aware of how settlement breeds scarcity, and scarcity, economic value. One example is found in William Cooper's short chapter, "Timber," in *A Guide in the Wilderness*, a chapter almost entirely devoted to explaining which of the various species of trees found in the Otsego Patent are best suited for fuel or lumber. *A Guide in the Wilderness* is composed as a response to several inquiries from a man named William Sampson about the area's prospects for settlement, and in answer to the question of whether "too great zeal for clearing may render it [the timber] in some time as scarce as it is now abundant?" (4), William writes one of the most powerful examples of how his speculative genius paints the area's environmental future:

> You ask whether there is no danger of a future scarcity of timber? I have always thought there was, and for these reasons: great part of our timber is not only of slow growth but when used for fencing, it is of short duration. Our winters require large supplies of fuel, and we have neither peat nor coal to resort to, when constant consumption in fuel and fencing will have rendered that most

[39] Ringe, *James Fenimore Cooper*, xii.

> necessary article scarce. Nor have we other mountainous lands to resort to for wood, nor stony ground for the construction of stone fences. The soil being all fit for culture, will be all cultivated, and the wood of course wasted. (28)

On its surface this conclusion to William Cooper's chapter on the ways to take advantage of the area's trees might seem to run counter to the capitalistic tone of the preceding discussion. Such a statement seems to clash with his more extensive lessons about which trees make for the best firewood or the most resilient roofs, but when considered within the proper context of the document itself—*A Guide in the Wilderness* is after all a work promoting the further settlement of the area—William's prophesy seems less a warning against overuse and more a recognition of an unavoidable consequence of the very settlement he advocates. A close reading of the final sentence makes clear that agriculture and harsh winters make deforestation inevitable: "The soil being all fit for culture, will be all cultivated, and the wood of course wasted."

For all of William Cooper's visionary gifts, he can offer no alternative future, no remedy to this dire forecast. He recognizes, perhaps dolefully, that where the land can be exploited it will be exploited. And he is not necessarily against the agricultural domination of the area. After all, the point of such land speculation was to make a fortune through the sale of the patent to settlers who wished to own fee simple farms. Still his tone is one of lament, as he anticipates the eventual deforestation that indeed did follow in the wake of the area's settlement. In *The Pioneers* his son offers an idealized magnification of William Cooper's environmental lament. Amplifying it into an uncommonly progressive conservationist ideology. Temple becomes a forceful advocate for the judicious use of Templeton's resources. That he happens to own a good amount of those resources, and that he views the waste of those resources as the destruction of his property, is no small factor in his reasoning. While the guarding of personal property is not as romantic an environmental philosophy as that held by Natty Bumppo, it at least represents one backed by law. Indeed at the core of this novel resides the conflict between those who claim legal ownership of the land and those who claim a natural right to live off of it. It is this conflict that separates Judge Temple and Natty Bumppo, men who both claim a land that is not really theirs and who both forward well-intentioned but divergent ideas about the best way to conserve that land's resources.

Conclusion

In both *The Pioneers* and William Cooper's *A Guide in the Wilderness*, the problem of environmental degradation is clearly one of European origin, of immigrants who in their rush to profit from the new American Eden, leave a trail of environmental degradation. In the final chapter of his book, entitled "Absurdities," William Cooper shares anecdotes about three naïve immigrants who came to the patent, and who, out of stubbornness made fools of themselves by assuming that

European methods of clearing the land of trees would work on the densely forested frontier of western New York. Among others, he chides a misguided Irishman who, through the use of "tackles, blocks, windlasses, and capstans, with other mechanical auxiliaries" tried to have the trees of his plot uprooted instead of cut down (46). The man, whose goal was to create a field free of stumps, quits after clearing only "a few acres at an enormous expense," (46) finding that the field of pits and compromised soil made for no less troublesome farming. "At length," writes William, "this gentleman found out that it was one thing to dress his pleasure grounds in Ireland, and another to clear the Wilderness in America" (46).

This and other examples make plain that while he at times reveals a glimmer of conservationist leanings, William is still very much a man concerned not with saving the land but with adopting the most efficient methods of profiting from it. William's speculative vision is nothing if not pragmatic, and he ends with a warning to others to adapt agricultural practices to this new land rather than the other way around; to learn from the example of the local farmers; and above all to be patient. "Gradual improvements," he says,

> are made, and will grow out of experience, and certainly I am not an enemy to ingenious speculations, nor even to theories, provided they be the result of long and attentive observation, and grounded upon facts well ascertained ... As yet I think it safer that the philosopher should learn from the farmer, than the farmer from the philosopher. (49)

The local farmer then, at least in William Cooper's estimation, becomes the best source for knowledge about the practical land management of the area. This overall belief in the agricultural destiny of the Otsego Patent reveals just how much William's environmental vision differs in essence from that of his son, who, in creating Judge Temple and Natty Bumppo presents two characters who despite their different worldviews still share a genuine concern for the land that extends beyond its agricultural potential.[40]

Critics like John Seelye persuasively argue that Judge Temple is—despite his conservationist leanings—still a "Republican visionary," who in his mind's eye sees in the wilderness a future of industrial and agricultural development and economic promise.[41] Seelye's reading of Natty's role in this first of the Leather-Stocking Tales is equally instructive, in that we cannot understand the old hunter unless we pay attention to how he is in some ways at odds with his origins in the Daniel Boone myth. "Cooper's version of Daniel Boone," writes Seelye, "is an aging hunter in retreat from civilization, not a harbinger of 'Refinement and

---

[40] For a thorough examination of early American farming and agricultural methods as related to the literary works of the eighteenth and nineteenth centuries see, Sweet, *American Georgics*, where he argues in part, "the literature of agrarian improvement developed a discourse of rural virtue that linked economic intensification ... to national political stability" (99).

[41] Seelye, *Beautiful Machine*, 352.

Enterprise' but a champion of wilderness ways who delivers scathing homilies about the wastefulness of the 'pioneers,' who gives Cooper's title an ironic dimension."[42] Yet in the end Natty surrenders to the onslaught of ignorant settlers with their equally ignorant despoliation of the land, and he once again retreats from the ever-advancing edge of civilization. Thus *The Pioneers* is perhaps one of the first American literary engagements with the Tragedy of the Commons. Temple's solution to the problem of conserving the non-exclusionary resource of western New York is to try to regulate it by limited-use agreements and hunting seasons. Natty, however, refuses to see himself as part of the problem at all, and refuses to agree to any social contract. So by the end of the book he must depart for another, as yet unspoiled, region to the west. As we will see in the following chapter, that story is told in Cooper's *The Prairie* (1827), a book that while quite different from *The Pioneers* stands as one of Cooper's most environmentally sophisticated novels.

---

42 Seelye, *Beautiful Machine*, 352–3.

Chapter 5
The Biogeography of *The Prairie*

"Then was your buffoon a fool to my Hector!"
—Natty Bumppo to Dr. Bat, *The Prairie* (1827)[1]

In the 14th installment of what would eventually become his 36-volume *Histoire naturelle* (1749–1788), the French naturalist George-Louis Leclerc, Comte de Buffon, managed to belittle the natural environments of both North and South America while also ridiculing the indigenous inhabitants of the two continents:

> In America ... animated Nature is weaker, less active, and more circumscribed in the variety of her productions; for we perceive, from the enumeration of the American animals, that the numbers of species is not only fewer, but, in general, that all the animals are much smaller than those of the Old Continent. Even those, which, from the kindly influences of another climate, have acquired their complete form and expansion, shrink and diminish under a niggardly sky and an unprolific land, thinly peopled with wandering savages. (5:129)[2]

It was quite an insult. Not satisfied with mischaracterizing the fauna of North America—a continent he had never visited—Buffon began to flesh out what he termed his theory of American degeneracy, an offshoot of his approach to natural history in which he argued that the American climate, lush forests, and water supply combined to create an environment prone to reducing the size and vitality of its inhabitants.

In this chapter I argue that James Fenimore Cooper—fully aware of how Buffon's derogatory ideas endured in the popular view—wrote *The Prairie* (1827) in part as a refutation of the Frenchman's theories. Even for Cooper, *The Prairie* is ambitious in scope. It includes among its events a prairie fire, a buffalo stampede, a murder mystery, battles between the Sioux and Pawnee, and one of the most disturbing scenes of frontier justice in all of American literature. Taken as a whole, however, *The Prairie* is a novel of ideas, investigating key social, political, religious, and cultural challenges facing the new Republic. At the core of these concepts is Cooper's imagined western American environment, exemplified by the uniquely American, and in 1827 still mysterious, Great Plains. As he had in *The Pioneers* (1823), Cooper pens a forward-thinking, if pessimistic, work of environmental fiction. His book laments the kind of seemingly inevitable destruction of the natural world that followed westward expansion and that

---

1 Cooper, *The Prairie*, 106. Subsequent references appear parenthetically in the text.

2 Buffon, *Natural History, General and Particular* [*Histoire naturelle*], 5:115. Subsequent references appear parenthetically in the text.

destroyed the very land it extoled. Specifically addressing Buffon at key points in the novel, Cooper simultaneously refutes and complicates the naturalist's portrayal of the North American environment, here represented by the bioregion of the Great Plains. He specifically chooses as his setting a harsh, dry land sharply at odds with the stifling swamps of Buffon's conjecture. Furthermore, Cooper employs the befuddled taxonomist Dr. Obed Bat (who has, in true Linnean fashion, Latinized his last name to Battius), paradoxically, as a critic of Buffon's ideas and as an exemplar of the very failings Cooper saw in the Eurocentric brand of Enlightenment science promoted by Buffon and his followers. Also, in rejecting Buffon's racial and national prejudices, Cooper constructs his own myth of the natural world of the United States by populating it, not with Buffon's diminutive weaklings, but with unnaturally large, and—especially in the case of American Indians—resourceful supermen. Ultimately, Cooper's rendering of America's biogeography and national character in *The Prairie*—a novel he started in New York and completed in Paris—functions as a rebuttal to Buffon's theory of American degeneracy and, as Cooper's own volume of natural and social history, represents an imagined American environment that, while no less fanciful than Buffon's, is far removed from the Frenchman's prejudiced depiction.[3]

Buffon's Theory, U.S. Culture, and the Great Plains

Discussing the last of his four "ingredients" for an environmental work of literature, Lawrence Buell writes that an environmental text should in part include "*some sense of the environment as a process rather than as a constant or a given*" and that "by this criterion, James Fenimore Cooper's *Pioneers* is a more faithful environmental text than the four ensuing Leatherstocking Tales because it never loses sight of the history of the community's development from wildness to town."[4] While *The Pioneers* is indeed remarkable in its early portrayal of the idea of bioregional succession from unaltered forest to human settlement, we would do well to also consider the environmental lessons that reside in *The Prairie*, especially when placed in the context of early Republic biogeographical discourse.

Explaining Buffon's ideas, Alan Bewell states that "even as the primary scientific focus of biogeography was on the natural or artificial distribution of plants and animals and their adaptation to new environments, its most controversial aspect lay in its reflection on the biological relationship between people and environments."[5] For Buffon and his later supporters, this ostensibly detrimental connection between the American climate and those who called it home explained the host of failings, biological and cultural, they saw in the new country. In the years following Buffon's publication of his theory of degeneracy,

---

[3] Elliott, Historical Introduction, *The Prairie*, xv–xvi.

[4] Buell, *The Environmental Imagination*, 8.

[5] Bewell, "Jefferson's Thermometer: Colonial Biogeographical Constructions of the Climate of America," 116.

U.S. thinkers, most famously Thomas Jefferson, attacked his ideas, dismantling his pseudoscientific representation of Americans and American nature. Jefferson, in *Notes on the State of Virginia* (1782–1787), rebuffed the theory with persuasive tables detailing the remarkable diversity of his country's wildlife, along with accurate information about the sizes and weights of American animals. Jefferson also refuted the idea that Native Americans were victims of degeneracy, arguing that any symptoms of this supposed condition were actually the effects of malnutrition and a grinding existence devoid of the luxuries of European life. He insisted that when offered the benefits enjoyed by Europeans, American Indian women were as capable of mothering children as European American women. Despite Jefferson's persuasive arguments and Buffon's own eventual abandonment of his idea of human degeneracy, the theory lingered in the popular consciousness for decades, proving a resilient, if illogical, way for those who despised America to criticize the fledgling nation. The persistence of Buffon's ideas was in part the result of the mass popularity of *Histoire naturelle.* A 1910 inventory of the sale catalogs for 500 private libraries put on the market in Paris from 1750 to 1780 found that Buffon's *Histoire naturelle* was the third most commonly owned work, with Pierre Bayle's *Dictionnaire historique et critique* at number one.[6]

While Buffon eventually backed away from some aspects of his theory, other European thinkers began to champion his earlier ideas. Figures such as Cornelius de Pauw and Guillaume Thomas Françoise Raynal promoted his theory as an explanation of what they considered a pervasive American biological and cultural degeneracy. "Buffon's analysis was pivotal," notes Philippe Roger, "All naturalist anti-Americanism can be traced back to it. It not only gave an account of America's general deficiency ... it also framed America's future, using the rhetorical model of decline."[7] Thomas K. Murphy writes that "in 1768, de Pauw excoriated the peoples of America, viewing them as little more than animals. De Pauw took a more radical stance on degeneracy than did Buffon, linking the New Continent not only with newness and inexperience, but with decay."[8] De Pauw was particularly abusive in his estimation of American Indian males, whom he viewed as "physically weak, lacking both sensibility and taste as well as a capacity for higher mental activity."[9] Connecting this idea of physical degeneracy with U.S. cultural output, Raynal declared, "One must be astonished that America has

---

6 Jefferson, *Notes on the State of Virginia*, 186; For a discussion of Buffon's life and contributions to science, see Jacques Roger's *Buffon: A Life in Natural History* (originally published in French in 1989). Cañizares-Esguerra, in *How to Write the History of the New World* (2001), provides an astute analysis of the philosophical, political, and historiographical implications of Buffon's theories; see also Fellows and Milliken, *Buffon,* 56.

7 Roger, *The American Enemy: The History of French Anti-Americanism*, 12–13.

8 Quoted in Murphy, *A Land without Castles: The Changing Image of America in Europe, 1780–1830*, 15.

9 Quoted in Murphy, *A Land without Castles: The Changing Image of America in Europe, 1780–1830*, 15.

not yet produced one good poet, one able mathematician, one man of genius in a single art or a single science." [10] Indeed, Raynal's Buffon-inspired charge of American cultural inferiority resembles the Reverend Sydney Smith's oft-cited argument that Americans

> have hitherto given no indications of genius. ... Their Franklins and Washingtons, and all the other sages and heroes of their revolution, were born and bred subjects of the King of England ... they have done absolutely nothing for the Sciences, for the Arts, for Literature, or even for the statesman-like studies of Politics or Political Economy. [11]

The fact that Smith wrote this criticism in 1820 speaks to the way Buffon's ideas spilled over from the late eighteenth century into the nineteenth.

Cooper probably became acquainted with Buffon's ideas early in his education, carrying these criticisms of his country with him into his writing. Although in 1826 the natural environment remained for the most part a sporadically and superficially treated topic in works of fiction by other US authors, Cooper had already published in his first Leather-Stocking Tales—*The Pioneers* (1823) and *The Last of the Mohicans* (1826)—two of the most environmentally conscious works of Early Republic fiction. Unlike the prior Leather-Stocking Tales, *The Prairie* takes as its setting a plains territory that illuminates the human presence on the land as well as the dangers of that environment. According to Donald Ringe,

> "the setting of *The Prairie* serves the perfect aesthetic function in the book. Unlike *The Pioneers*, *The Prairie* depicts a world man cannot pretend to master, for it is completely beyond his control; and unlike *The Last of the Mohicans*, it presents a gaunt, bare, hostile nature that leaves man naked to the elements. [12]

As Ringe indicates, there are key environmental differences between *The Prairie* and the first two Leather-Stocking Tales, and while *The Pioneers* forwards a conservationist ethic, *The Prairie* explores a more modern conception of environmentalism, especially in its portrayal of the Great Plains as a wild land that, unlike the New York scenes of *The Pioneers*, does not readily supply human needs.

Thus the exotic plains setting held literary promise, especially for a book that Cooper envisioned at the time as the last of his Leather-Stocking series, and in this novel Cooper replied to those European critics enamored with Buffon's ideas, perhaps the most troublesome being his contention that the ostensibly stunted character of American mammals derived from an unfavorable natural

---

10 Quoted in Golden and Golden, *Thomas Jefferson and the Rhetoric of Virtue*, 357.

11 Sydney Smith, *The Works of the Rev. Sydney Smith*, 141.

12 Ringe, *James Fenimore Cooper*, 46.

environment.[13] In particular, Buffon portrayed America's climate, water resources, vegetation, and apparent newness as a continent as the causes of degeneracy:

> In the New Continent, there are more running waters, in proportion to the extent of territory, than in the Old; and this quantity of water is greatly increased for want of proper drains or outlets ... [A]s the earth is every where covered with trees, shrubs, and gross herbage, it never dries. The transpiration of so many vegetables, pressed close together, produces immense quantities of moist and noxious exhalations. In these melancholy regions, Nature remains concealed under her old garments, and never exhibits herself in fresh attire; being neither cherished nor cultivated by man, she never opens her fruitful and beneficent womb. Here the Earth never saw her surface adorned with those rich crops, which demonstrate her fecundity, and constitute the opulence of polished nations. In this abandoned condition, every thing languishes, corrupts, and proves abortive. The air and the earth, overloaded with humid and noxious vapours, are unable either to purify themselves, or to profit by the influences of the Sun, who darts in vain his most enlivening rays upon this frigid mass, which is not in a condition to make suitable returns to his ardour. (5:135–6)

In Buffon's anthropomorphized portrait, American nature is a sterile old maid jilted by the agricultural industry of mankind, draped in layers of rotting vegetation, and abandoned in a miasma of decay. Buffon portrays the American environment as a stunted world that fails to fulfill what Annette Kolodny terms "America's oldest and most cherished fantasy: a daily reality of harmony between man and nature based on an experience of the land as essentially feminine—that is not simply the land as mother, but the land as woman."[14] While Buffon's representation takes the guise of a failed, feminized land, Cooper, himself no stranger to feminizing the North American natural world, takes a different approach by presenting an arid prairie realm that acts as the antithesis of Buffon's watery, debilitating landscape. In his introduction to *The Prairie*, Cooper writes that "American prairies are of two kinds" (4). Of those plains east of the Mississippi, Cooper asserts that while they "are fast becoming settled . . . [t]hey labour under the disadvantages of scarcity of wood and water,—evils of a serious character, until art has had time to supply the deficiencies of nature" (4). In a statement that resembles Buffon's ideas about the ability of human alterations to improve the land, Cooper maintains that the eastern plains are "susceptible of high cultivation" and that "the enterprise of the emigrants is gradually prevailing against these difficulties" (4). With these sentiments, he betrays his own Enlightenment-era justification for the human alteration of the landscape, arguing that "enterprise" will purge the landscape of the "evil" of failing to supply human comforts.

---

[13] For a detailed, if dated, study of the cultural context, composition, and reception of *The Prairie*, see Orm Överland's *The Making and Meaning of an American Classic: James Fenimore Cooper's "The Prairie."*

[14] Kolodny, *The Lay of the Land*, 4.

Cooper's description of the second "kind" of American plains environment, however, the "Great Prairies" stretching west of the Mississippi, takes on a different tone. He argues that these prairies are "a vast country, incapable of sustaining a dense population. ... Rivers abound, it is true; but this region is nearly destitute of brooks and the smaller water courses, which tend so much to comfort and fertility" (4). As this description reveals, Cooper's fictional Great Plains landscape is as infertile and inhospitable as Buffon's, but the region has shifted from a "frigid mass" (Buffon 5:136) to a "comparative desert" that "interpose[s] a barrier to the progress of the American people westward" (Cooper 4). Cooper's prairie landscape does have pockets of relief from the desolate grassland, but these oases are rare and known only to those with eyes acclimated to the subtle variations in the land. For instance, upon their first meeting, Ishmael Bush asks Natty where they may find "sweet water, and a good browse for the cattle" (17). Natty leads them from the waterless plain to a place where

> "a clear and gurgling spring burst out of the side of the declivity, and joining its waters to those of other similar little fountains[,] ... their united contributions formed a run, which was easily to be traced, for miles along the Prairie, by the scattering foliage and verdure which occasionally grew within the influence of its moisture. (18)

While the environment of Cooper's novel is not devoid of water, it is a scarce resource that scrawls thin veins of lushness across the land. It is important to remember that like Buffon, Cooper had never actually visited the Great Plains. In writing *The Prairie* he relied on his own imagination and on the descriptions of explorers, especially Stephen H. Long, whose early travels in the Great Plains were related in Edwin James's *Account of an Expedition from Pittsburgh to the Rocky Mountains under the Command of Major Stephen H. Long* (1823), a book that served as a major source for the novel.[15] Thus, while Cooper's representation of the landscape is better informed than Buffon's, it is no less a combination of second-hand accounts and flights of fancy, though, unlike Buffon, Cooper never claimed his book as anything more than a work of fiction.

These two imaginary American environments diverge most radically in how they should be viewed by potential settlers. Buffon's America, devoid of any environmental variety, is a series of "melancholy regions" best avoided—if not for aesthetic reasons then for the potential degeneracy accompanying prolonged contact with the area's "noxious exhalations." Cooper's "Great American Prairies," on the other hand, "form one of nature's most majestic mysteries," and while Buffon is quick to paint the whole of America as a chilly swamp, Cooper is more attuned to the continent's diverse bioregions and climates (4).

---

[15] Elliott, Historical Introduction, *The Prairie*, xvii.

Vespertilio; Horribilis, Americanus

Like his descriptions of the North American prairie, Cooper's portrayal of Dr. Bat works as a further reaction against Buffon, but his characterization of Bat is something of a paradox. Cooper's comic naturalist is both a satire of those scientists who advocated the theory of American degeneracy and a vocal critic of Buffon in particular. In one particularly comic passage, Bat makes his way back to camp with a harrowing tale. Finding only Ellen Wade awake, he relates his discovery of a remarkable new species, proclaiming that his "name will go down to posterity with that of Buffon—Buffon! a mere compiler! one who flourishes on the foundation of other men's labours" (69). "Listen, girl," Bat says, "and you shall hear, with what a treasure it has been my happy lot to enrich the pages of Natural History" (70). Reading from his field notes, Bat details, in true taxonomic fashion, the fantastic creature he met in the night: "*Vespertilio; Horribilis, Americanus.* ... *Greatest length* eleven feet, *height*, six feet. ...*[T]eeth*, serrated and abundant. ... *Talons*, long, arquated, dangerous. *Ears*, inconspicuous. *Horns*, elongated, diverging and formidable. ... *Habits*, gregarious, carnivorous, fierce, and fearless" (71). Bat exclaims that this monstrous creature "is an animal, which will be likely to dispute with the Lion, his title to be called the King of the Beasts!" (71). Upon finishing his description of this newly named prairie monster, he recounts, "I was pursued—hunted—and in a danger that I scorn to dwell on—what's that" (72). Recognizing the outline of his terrible "*Vespertilio; Horribilis, Americanus*," the naturalist cries, "It comes! it comes!" (72). As the beast approaches, it lets out "a roar, or rather a shriek," which is followed "by an uncontrolled fit of merriment from the more musical voice of Ellen," who, regaining her composure, says, "It is your own ass! ... Do you refuse to know an animal that has laboured so long in your service!" (72). Through such comic scenes, Cooper derides the disciples of Buffon and Linnaeus for their fragmentary and short-sighted methods of knowing the natural world, and he also subtly ridicules his own fact-gathering methods when he has Bat condemn Buffon as "a mere compiler! one who flourishes on the foundation of other men's labours" (69). Here Cooper, well aware of his own reliance on the written accounts of explorers for his own understanding of the Great Plains, engages in a playful moment of self-mockery.

Bat's ineptitude may seem at first the work of an author hostile to the principles and achievements of science, as well as the Enlightenment in general, but Cooper's opinion of reason was more nuanced than such passages may initially indicate. Harry Hayden Clark characterizes the balance the religious Cooper struck in his view of scientific thought, writing that

> Cooper was persistent in admitting the efficacy of science in meliorating the condition of mankind. He was equally persistent in refusing to admit that science was the key to man's universal emancipation. Placed in its proper position and properly subordinated, science was a useful tool; but it was not the be all and end all of human accomplishment.[16]

[16] Clark, "Fenimore Cooper and Science," 194.

Bat, argues Clark, represents the dangers of trusting scientific rationalism to solve larger philosophical and religious problems and he adds that Cooper employs the frequently baffled naturalist "who studies Buffon and Linnaeus, as a means of satirizing the extreme view that science alone has the power to eradicate all evil from the world."[17] Through his bumbling scientist, Cooper asserts that American nature is not fully classifiable or understandable by established European systems of natural history.

While Dr. Bat is perhaps one of the first of his kind in U.S. literature, he actually originates from a long literary tradition of comical academicians, such as those found in Ben Jonson's *Alchemist* (1610) and Jonathan Swift's *Gulliver's Travels* (1726).[18] Bat, in particular, represents how taxonomic systems of compartmentalizing and naming species fail to account for the ways that organisms interact to form multifaceted communities of interdependence. With the significant exception of William Bartram, biologists of the late eighteenth and early nineteenth centuries tended to eschew a larger ecological approach in favor of imposing a strict Linnaean taxonomic system of classification and inventory. These systems invite, especially in the case of plants, a mutilated portrayal of the organism in question. Rather than a representation of the entire plant or ecosystem, the taxonomist offers cold descriptions and drawings of a dissected organism—flowers, leaves, and stems. Rarely does this taxonomic system afford a view of the entire plant, and only in the most cursory way do the descriptions of the plants mention their interactions with the environment in which they are quite literally rooted. Like other Romantics, such as Goethe, Cooper took issue with this type of reductionism, as found in Buffon and in Linneaus. In a manner similar to that of the actual naturalists of the day, Cooper's Bat takes an extremely fragmented approach in his encounter with the donkey, building his conception of the creature from his observation of individual parts rather than of the whole. Bat's narrow scientific method combined with his tendency to sensationalize the quotidian creatures he encounters leads to his creation of an imaginary monster out of a friendly donkey.

It would seem from his descriptions of nature that rather than supporting the idea of American biological degeneration, Bat promotes a humorously misguided form of amplification or enhancement in which common donkeys become horrible, horned terrors. For instance, in another scene Natty tests Bat's spurious scientific method by pitting his dog Hector's nose against the skills of the naturalist, asking the man to name the unseen "creatur'" that has been "cracking the twigs" in a nearby brake (105). Bat says, "It exceeds the limits of earthly knowledge! Buffon himself could not tell whether the animal was a quadruped, or of the *order*, serpens! a sheep, or a tiger!" (105). Natty says, "Then was your buffoon a fool to my Hector!"

---

17 Clark, "Fenimore Cooper and Science," 195.

18 Scientists, serious and comic, are well represented in Cooper's fiction, with perhaps his most satirical being Dr. Reasono of *The Monikins* (1835), who is literally a talking monkey.

and quickly discovers from the dog's body language that the unseen animal is none other than the very human Duncan Uncas Middleton (106). In each case Cooper positions the comical Bat as the personification of the tendency of scientists such as Buffon to adopt ill-informed and fanciful interpretations of common aspects of the American environment. As his various moments of misinterpretation indicate, Bat views American nature through a distorted lens, one in which the American landscape becomes not only enlarged but also grotesque. This tendency is integral to his character, and Cooper even alludes to it when Bat relates his initial scrape with *Vespertilio; Horribilis, Americanus*. Ellen, teasing the dimwitted man asks, "Is it then a creature of your forming," to which Bat replies, "Is the power to give life to inanimate matter the gift of man; I would it were! You should speedily see a Historia Naturalis, Americana, that would put the sneering imitators of the Frenchman de Buffon to shame!" (70). Bat then follows with a description of how he would improve the physiology of "all quadrupeds" by replacing their limbs with levers and wheels, even suggesting that the animals should exude an organic lubricant to "assist in overcoming the friction" (70, 71). Bat much prefers his own twisted vision of nature to the reality around him, and at those times when he does recognize creatures for what they are, he egotistically suggests improvements on their anatomy.

In another of Bat's cases of failed identification, he describes the hiding Pawnee warrior, Hard-Heart as "a basilisk! ... An animal of the *order* serpens[;] ... a monster that nature delighted to form" (182). In mistaking the Apollo-like Hard-Heart for a lizard, Bat figuratively degenerates the man, lowering him from human to reptile, similar to how Buffon (before eventually backing away) denigrates and dehumanizes American Indians by describing them as "brutish men" characterized by "stupidity or indolence" who "devour every thing, and propagate nothing."[19] Bat, in trading Hard-Heart's human appearance for that of a lizard, follows suit, and although Natty insists that the figure is a hiding person, Bat maintains that "the animal is not even of the *class* Mammalia, much less a man" (184). Bat's choice of a large lizard is doubly significant when taken into consideration with Buffon's contention that the only types of animals that fare well in the otherwise debilitating North American climate are insects and reptiles. Buffon declares that "the toads, the frogs, and other animals of this kind, are ... very large in America" and the continent is "limited to the production of moist plants, reptiles, and insects and can afford nourishment only to cold men and feeble animals" (5:132, 136). Bat—who happens to carry a pistol loaded with "minute particles of lead, adapted to the destruction of the larger insects and reptiles"—imaginatively transforms Hard-Heart into an example of the type of animal best matching Buffon's theories, and he does so in a comically exaggerated way, positing the existence of man-sized lizards on the Great Plains (70). Scenes such as these reveal how Cooper attacks

[19] Quoted in Bewell, "Jefferson's Thermometer: Colonial Biogeographical Constructions of the Climate of America," 128).

Buffon's ideas by calculated, satirical overstatement rather than, as Jefferson had, by offering logical, empirical responses.

A closer look at Cooper's dimwitted taxonomist reveals a somewhat ominous facet of his character. As critics such as Clark, Ringe, and Anne Perrin have argued, Bat serves as a straw man for Cooper's criticism of certain Enlightenment ideas. In revealing his personal scientific philosophy, one apparently matching that of scientists such as Buffon, Dr. Bat claims that in the future, science will allow humans to "become the Master of all learning, and consequently equal to the great moving principle" (180). Ringe maintains that in this belief, the otherwise innocuous Bat assumes a sinister aspect because, "though certainly superior to Bush in education and refinement of character, Dr. Bat, whose name suggests his intellectual blindness, is curiously like him in assertion of his own ego and in the assumption that his mind and will represent the standard for judging the universe." Mirroring Ringe's reading of Bat, Perrin argues that "it is with Battius that Cooper constructs the most extended debate regarding the values of nature vs. those of civilization," and that through Bat, "Cooper takes to task enlightened scientific thought."[20] Bat's ludicrous misinterpretations of the natural world prove him a fool bound by misinformed scientific methods and faulty powers of observation, but his unwavering belief in the primacy of his own obviously flawed ideas makes him a menacing figure. Bat serves as Cooper's warning that self-serving foolishness masquerading as the voice of reason is a danger to any society unwilling to question its validity.

Cooper's "Sturdy Men"

In the fifth volume of *Histoire naturelle*, Buffon elaborates on his view that the American landscape fosters a general stunting of animals and people by alleging that "no American animal can be compared with the elephant, the rhinoceros, the hippopotamus, the dromedary, the camelopard, the buffalo, the lion, the tiger, &c." (5:115). It would seem that this idea, in particular, drew the ire of Jefferson, who directly addresses it in *Notes on the State of Virginia*, where he attempts to refute Buffon by referencing the fossilized remains of mammoths that had turned up in recent years and also by citing tales hinting that the elephantine animals still roamed along the shadowy corners of the American map. On February 24, 1803, Jefferson wrote to the French naturalist Bernard Lacépède (who, following Buffon's death, would publish the final eight volumes of *Histoire naturelle*) about the impending Lewis and Clark expedition, explaining that one of its goals was to gather evidence of American megafauna, including the mammoth and the giant ground sloth, or Megatherium. He relates,

[20] Ringe, *James Fenimore Cooper*, 47; Perrin, "Opened Frontiers, Closed Deserts: The Contradictions between Source and Text in James Fenimore Cooper's *The Prairie*," 74.

> We are now actually sending off a small party to explore the Missouri to it's [*sic*] source, and whatever other river, heading nearest with that, runs into the Western ocean. ... It is not improbable that this voyage of discovery will procure us further information of the Mammoth, & of the Megatherium.[21]

Jefferson's interest in these enormous creatures was understandable; the discovery of a living mammoth would have served as the ultimate proof that American animals were far from "degenerate." In fact, in a letter to Buffon accompanying a gift donated to the French Cabinet du Roi, or Royal Cabinet of Curiosities, Jefferson wrote, "I am happy to be able to present to you at this moment the bones & skin of a Moose, the horns of the Caribou, the elk, the deer, the spiked horned buck, & the Roebuck of America."[22] Jefferson perhaps saw the selection of specimens as further proof against the Frenchman's idea that American animals were small in size.

Cooper may have been aware of how Buffon's theories had irritated Jefferson; he, after all, begins *The Prairie* with the opening of the land acquired under the Louisiana Purchase, and, as Orm Överland states, "the first pages of the novel constitute in effect an essay on the Louisiana Purchase and its consequences."[23] Cooper sets the action in 1804, the very year of the opening of the land to exploration, the same year that Jefferson's "Corps of Discovery," led by Lewis and Clark, departed St. Louis. He even has Natty returning from the West Coast, a journey of 10 years from which the trapper learns that the landscape is no cold swamp of Buffon's conjecture, but rather a "weary path" in which one could "travel weeks," and still see the same "barren belt of Prairie" (23, 24, 24). While the Lewis and Clark expedition found no mammoths lumbering alongside the grazing buffaloes of the plain, Cooper does include an intriguing substitute for those large mammals—Ishmael Bush and his sons. In a gesture that echoes Bat's hyperbolic natural descriptions of mundane organisms, Cooper describes these "sturdy men" as almost mythic in proportion: Asa Bush suggests the bearing of a "giant," and his father, despite his years, displays the "unwieldy, but terrible, strength of the elephant" and moves "with the strides of a giant" (11, 12, 363). Unlike Bat's descriptions of giant lizards, Cooper's squatters live up to their portrayal as a family of giants, indicating that the Bushes represent the opposite of Buffon's theory of human degeneracy, at least in terms of physical prowess.

---

21 Jefferson to Bernard Lacépède, *In Letters of the Lewis and Clark Expedition with Related Documents, 1783–1854*, 15–16.

22 Jefferson to Buffon, October 1, 1787, in *Writings*, 909–10. Andrew Lewis points out that along with such transatlantic exchanges of specimens, there was also a domestic demand for certain, profitable organisms. For example, at least one entrepreneurial western Pennsylvania farmer wrote to Benjamin Smith Barton, a botanist at the University of Pennsylvania, requesting poppy seeds and cultivation instructions for the making of opium (67).

23 Överland, *James Fenimore Cooper's* The Prairie*: The Making and Meaning of an American Classic*, 147.

After Natty leads the Bushes to a suitable spot to set up camp, the squatters begin, like locusts, to take their toll on the environment. The "giant" Asa starts felling trees for the construction of a temporary settlement, and Natty looks on: "As tree after tree came whistling down, he cast his eyes upward, at the vacancies they left in the heavens, with a melancholy gaze, and finally turned away, muttering to himself with a bitter smile, like one who disdained giving a more audible utterance to his discontent" (19). Natty, weathered from a life on the ever-shifting American frontier, views the felling of the trees as the very condition he has tried to escape, and his complicity in the act by leading the Bushes to the spot perhaps contributes to his "bitter smile." In *The Prairie*, he again witnesses the human blade of settlement cutting swaths of homesteads, farmland, towns, and outposts ever westward, and it is through Natty that Cooper champions a more holistic appreciation of the environment. Though brutish outcasts, the Bush family, led by Ishmael, possesses their own brand of environmental consciousness in that by the end of the book they know enough to abandon their plan to settle in the plains. In his introduction to the novel, Cooper makes it clear that he views the Great Plains as inhospitable to all but Native American settlement, and having failed in their attempt, the Bush family (in a reversal of Natty's westward tendency) heads back to the eastern settlements. Ultimately, the Great Plains are too harsh even for these almost mythically strong European American settlers, a situation that flies in the face of both Buffon's idea of degeneration and of Cooper's other frontier novels in which the land must always succumb to European designs.

As Bewell relates, Buffon expressed a different view, arguing that "the inferiority of the American environment and its plants and animals could be seen as a symptom of the failure of its indigenous people to actively modify it to meet their needs."[24] According to Buffon's theory, fruitful development and cultivation of the soil would lead to a positive transformation in the general climate, one that would in turn strengthen the animals, plants, and even humans who resided there. He insisted that the American Indians had failed to improve the land and that it had taken a toll on their bodies, their minds, and their ability to reproduce, proclaiming that

> the American savage be nearly of the same stature with men in polished societies, yet this is not a sufficient exception to the general contraction of animated Nature throughout the whole Continent. In the savage, the organs of generation are small and feeble. He has no hair, no beard, no ardour for the female. Though nimbler than the European, because more accustomed to running, his strength is not so great. His sensations are less acute; and yet he is more timid and cowardly. (5:129)

Cooper addresses this argument in *The Prairie* when Natty asks Bat his opinion of how the Native Americans have made use of their environment. In answering,

[24] Bewell, "Jefferson's Thermometer: Colonial Biogeographical Constructions of the Climate of America," 120.

Bat echoes Buffon's view of American Indians as childlike beings who have failed to improve themselves and, in turn, their world. After Bat cites the Egyptian pyramids as monuments to a culture's past adulthood and that country's "present condition of second child-hood," Natty asks him, "And what see you in all this?" to which the naturalist replies, "A demonstration of my Problem, that nature did not make so vast a region to lie so many ages an uninhabited waste" (239). Put another way, the barren prairie stands as proof of the degeneracy of the American Indians, a race yet to reach the same cultural adulthood the Egyptians attained, and lost, centuries ago.

Natty and the naturalist follow this initial exchange with a discussion of the role of morality in the rise and fall of human societies, in which Natty does not so much refute Bat's ideas as find a way to apply those concepts to the North American prairie. In summing up this conversation, Lance Schachterle notes that "characteristically speaking at cross purposes with Leather-Stocking, Bat, by praising his own sacrifices in the name of science, initiates a discussion which evolves into an extended fantasia on lost Indian empires on the prairie."[25] Natty contributes to this "fantasia" by building on Bat's example of the Egyptians and forwarding a theory of his own: America is no new world. He explains that if societies are destined by human evil to fall, then there is "reason to think that he has wrought his evil here, as in the Countries you call so old. Look about you, man; where are the multitudes that once peopled these Prairies; the Kings, and the Palaces; the riches and the riotousness, of this desert?" (240). Bat asks "Where are the monuments, that would prove the truth of so vague a theory?" To this Natty answers, "They are gone. Time has lasted too long for them[.] ... This very spot of reeds and grass on which you now sit, may, once have been the garden of some mighty King. It is the fate of all things, to ripen, and then to decay" (240–241). Throughout this exchange Natty applies Buffon's ideas about degeneracy evenly across continents, arguing for a cyclical understanding of social development and cultural decomposition, thus arguing that America is no young world, but rather an exceedingly old one in which even its ruins of former glory have crumbled into dust.

Cooper, by attempting to rehabilitate the image of the American Indian, contends that the American prairie is not Buffon's infantile world. In crafting Indians who are far from Buffon's "timid" beings, Cooper directly counters these derogatory views. Just as the Bushes become elephantine in strength, so too do Cooper's American Indians become cunning, perceptive, and physically powerful beyond all reality. For example, in *The Prairie*'s first portrayal of an American Indian, Cooper describes the Sioux chief, Mahtoree, as "a warrior of powerful frame" and "an athletic and dark-looking" man of a "dangerous race" (45, 38, 40). This portrait stands in strong opposition to Buffon's description of American Indians as impotent cowards. Jonathan Arac argues that in *The Prairie*, Cooper "again sets 'good' Indian against 'bad' Indian" and that his "Indians do no more

---

25 Schachterle, "On *The Prairie*," 126.

than provide occasions for the various contrasting and conflicted relations among the whites," but this generalization overlooks how these indigenous characters work to refute earlier, racist European and Euro-American portrayals of Native Americans.[26] Wayne Franklin points to Cooper's acquaintance with the Skidi Pawnee statesman Petalesharo as a strong influence on *The Prairie*, explaining that Cooper so admired the man that he took him as the model for Hard Heart.[27] Cooper even found support for his high opinion of Petalesharo in another key textual source for *The Prairie*: Edwin James's *Account of an Expedition from Pittsburgh to the Rocky Mountains under the Command of Major Stephen H. Long* (1823). In this book, James details the North American prairie landscape, and he also describes Petalesharo as possessing "the finest form, tall, muscular, exceedingly graceful, and of a most prepossessing countenance."[28] By the time Cooper began work on *The Prairie*, he must have known that a figure like Petalesharo represented a living rebuttal to Buffon's portrayal of American Indians.

Cooper learned about American Indians from written accounts and from those Native American figures he met in the early 1820s. He possibly met with members of tribal delegations traveling to Washington, DC, and he certainly learned about Native Americans from the Moravian missionary John Heckewelder's *An Account of the History, Manners, and Customs, of the Indian Nations, Who Once Inhabited Pennsylvania and the Neighboring States* (1818). As Wayne Franklin suggests, it is also possible Cooper read Heckewelder's second book, *A Narrative of the Missions of the United Brethren among the Delaware and Mohegan Indians* (1820), along with John Long's *Voyages and Travels of an Indian Interpreter and Trader* (1791) and Alexander Henry's *Travels and Adventures in Canada and the Indian Territories* (1809) (473–4). Franklin also suggests that there is a high probability that in 1821 Cooper briefly met with the principle chief of the Omaha, Ongpatonga, at the same time that he encountered the prominent Skidi Pawnee statesman, Petalesharo (477–80). While *The Prairie*'s impossibly formidable Indians, like Buffon's Native American weaklings, are indeed cartoonish figures too often labeled as "good" and "bad," Cooper does imbue his American Indians with honor and humanity, something missing not only from the works of Buffon, Raynal, and de Pauw, but also from too many other contemporary literary and scientific representations. Cooper's presentation served as a better option than the hateful literary portrayals of Native Americans by his predecessor, Charles Brockden Brown with *Edgar Huntly* (1799), his contemporary Robert Montgomery Bird with *Nick of the Woods* (1837), and his successor Mark Twain with *Huck Finn and Tom Sawyer among the Indians* (1884).

---

26 Arac, *The Emergence of American Literary Narrative, 1820–1860*, 10.

27 Franklin, *James Fenimore Cooper: The Early Years*, 478.

28 Quoted in Franklin, *James Fenimore Cooper: The Early Years*, 479.

Conclusion

As *The Prairie* suggests, Cooper's nation is an amalgam of identities, of environmental wealth and abuse, and—in the case of injustices against American Indians—of deep-cutting wrongs that scar the social face of the country, degrading the human environment just as clear-cutting and other wounds of settlement mar the land. Cooper's refutation of Buffon in *The Prairie* was just one method of combating what he saw as unjust criticisms of his country. Though he would in later years adopt a pronounced critical view toward his own nation, in the years immediately following the publication of *The Prairie*—that is, during his seven-year stay in Europe—Cooper took pains to defend the United States against foreign critics. In addition to what he viewed as abusive portrayals of Americans published in the European press, Cooper also took issue with Fanny Trollope's satirical jabs at America in her travel book, *Domestic Manners of the Americans* (1832). Feeling it his duty to respond to these criticisms, Cooper addressed national issues in politically charged works such as his short satire "Point de Bateaux à Vapeur—Une Vision" (No steamboats—A vision), which he published in French in an 1832 issue of the Paris journal, *Paris, ou le livre des cent-et-un* (Paris, or the book of one hundred and one) and directed at what he viewed as ill-informed European critics who ridiculed a country they did not understand. Even years after the publication of *The Prairie*, Cooper still felt the need to address Buffon's theory of American degeneracy. In one passage from his 1838 travelogue *Gleanings in Europe: Italy*, he jokes, "This night I first learned to respect a musquito. If Buffon had in view the comparative merits of these insects when he broached his theory of the inferiority of the animal nature of America to that of Europe, there is more apology for the extravagance of the supposition than is commonly thought among ourselves."[29] Cooper was plainly concerned with how literary and scientific constructions of the American natural world influenced the reputation of the United States both at home and abroad, and for him the American environment was inseparable from the development of a multifaceted and at times unsavory American identity.

[29] Cooper, *Gleanings in Europe: Italy*, 17.

Chapter 6
Envisioning Disaster

It is dark disaster that brings the light.
—Maurice Blanchot, *The Writing of the Disaster* (1980)[1]

"Enlightenment philosophers ... dismissed the supernatural forces that dwarfed all efforts to contain their fury. They argued instead that disasters were acts neither of God nor—entirely—of nature."
—Marie-Hélène Huet, *The Culture of Disaster* (2012)[2]

When asked during an interview what he would take with him if his house were burning, the French artist Jean Cocteau said, "I'd take the fire."[3] If we, for a moment, allow his remark a small measure of seriousness, we realize that the more compelling rationale for taking the fire is not to save the house, but to save the fire, to control the uncontrollable—to hold the flames in our hands. Literary representations of catastrophe gesture toward this otherwise impossible preservation of violence incarnate. They capture that which otherwise defies containment. And if literary portrayals of natural disasters amount to a taking of the fire from the burning house of human experience, what then should we take from the fire itself?

Early American environmental fiction presents an instructive set of early texts with which to reconsider the cultural meaning of natural disasters because they predate so many of the types of anthropogenic, or at least human-exacerbated, disasters that threaten us now. These early texts come from a time in which the largely unknowable whims of weather and geologic forces stood as the most cataclysmic and unstoppable of threats, and also at a time in which those disasters were not as severely worsened by the human modifications to the landscape. Just as Cooper's portrayal in *The Pioneers* of the migration of a flock of passenger pigeons functions as a form of literary time capsule, so too do the glimpses of natural disasters in early texts offer a glimpse of how we imagined such catastrophes before we became so adept at spoiling our own nest.

Many of the words we use to communicate disaster are inextricably linked to the ruinous workings of the natural world. As Marie-Hélène Huet has noted, "Calamity" originates from the Latin for "stalk," an apt emblem for that which was repeatedly threatened by famine or some other agricultural peril. The word "peril" is derived from a convoluted Latin etymology thought to originally convey

---

1 Blanchot, *The Writing of the Disaster*, 7.

2 Huet, *The Culture of Disaster*, 6.

3 Cocteau, *Professional Secrets*, 249.

the possibility of losing a ship (and its cargo) to one of the various risks of ocean travel. "[R]isk itself," writes, Huet, "may have come from the Latin *resecum*, that which cuts, hence the reef that can sink a ship." And "disaster" refers to the traveler's fear of losing a guiding star, "of having been disowned by the stars that ensure a safe passage through life."[4] Tracing the etymology of these words shows us how our ancestors suffered at nature's frustratingly non-sentient activities. Even today it is easy to understand the impulse to make sense of such destruction by blaming gods and a personified nature alike. Yet as the two epigraphs that open this chapter suggest, disasters bring more than darkness, and none are without their human component.

In considering the question of how early national writers and artists portrayed natural disasters, I examine two novels relating profound and instructive literary representations of natural calamity: James Kirke Paulding's *Westward Ho!* (1832) and James Fenimore Cooper's *The Crater; or, Vulcan's Peak* (1847). I place these literary portrayals in conversation with the sometimes equally cataclysmic landscape paintings of Thomas Cole. While none of these texts present what we might today recognize as expressly didactic environmental cautionary tales, all of these works, especially *The Crater* and Cole's paintings of storms raging over human-altered landscapes, take a sophisticated allegorical stance on the problems of the human relationship with the land.

"The disquieting fact," writes Glen A. Love, "is that we have grown inured to the bad news of human and natural disasters. The catalog of actual and potential environmental crises is by now familiar to us all, so familiar as to have become dismissible."[5] In the nineteenth century, human destruction of the environment could rarely be as powerfully realized as the innate and unstoppable force of a volcano, earthquake, or tornado. This is not to say that these authors and artists did not recognize the amalgamation of human interests in the natural disasters they portray. What they do offer in their works is a representation of the violence of the natural world as an instructive and essential facet of nature. Even when presented allegorically, such as in Cooper's *The Crater*, these narratives function in an intellectual and instructive manner as cautionary tales that challenge us to learn from our mistakes.

Reimagining the New Madrid Earthquakes

> The rivers they boiled like a pot over coals,
> And mortals fell prostrate, and prayed for their souls;
> Every rock on our borders cracked, quivered, and shrunk,
> And Nackitosh tumbled, and New Madrid sunk.[6]

---

4 Huet, *The Culture of Disaster*, 3.

5 Love, *Practical Ecocriticism*, 14.

6 Schoolcraft, *Transallegania; or, the Groans of Missouri*, 23.

These lines are from Henry Rowe Schoolcraft's *Transallegania; or, the Groans of Missouri*, an 1820 epic poem that in part reimagines one of the most violent, and, ironically, least fatal natural disasters to strike the southern portion of the United States: the New Madrid Earthquakes. This series of quakes that struck the Central Mississippi Valley during the winter of 1811 and 1812 was the most powerful in recorded North American history. People as far away as Quebec felt the tremors, and for those few settlers, travelers, and American Indians who experienced the quake firsthand, it was nothing short of apocalyptic. During the strongest of the quakes, large swaths of land sank or rose as much as twenty feet; streaks of lightning ran along the ground; fissures opened in the earth spewing forth sand and water; new landforms such as the 18,000-acre Reelfoot Lake were born in a matter of minutes. The Mississippi was altered to such an extent that experienced captains had to relearn the river. In all, 150,000 acres of forest and prairie burned.[7]

This calamity—whose human impact was muted by its remoteness from larger settlements—created surprisingly few literary aftershocks. Aside from some brief accounts by travel writers and natural historians such John James Audubon, Thomas Nuttall, and Timothy Flint, the event was apparently treated only in Schoolcraft's poem and in one dramatic chapter from James Kirke Paulding's 1832 novel, *Westward Ho!* Paulding's novel in particular complicates the cultural and geopolitical *placement* of the New Madrid area, which, at the time of the earthquakes, was essentially a geopolitical void.[8]

In his study of the role of catastrophe in the stormy paintings of J.M.W. Turner, the philosopher James Williams invokes Gilles Deleuze's ideas on the cultural effect of disasters, writing that Deleuze offers "a world-view where catastrophe plays a positive role" and that "geological catastrophism is expressed through the ever-present processes of deterritorialisation and reterritorialisation; no well-defined territory is safe or static: it is constantly undone and remade."[9] But what are we to make of the role of a catastrophe in an ill-defined territory, specifically 1811 Missouri Territory? And how might such a disaster influence the identity of the larger region from which it originates? As Williams asserts, for Deleuze, a "catastrophe is only a stage that allows us to express the differential changes that underlie identity in a world committed to identity and representation."[10] Such a

7 Penrick, *The New Madrid Earthquakes*, 99.

8 It is worth noting the degree to which Cooper and Paulding were mutual influences. Paulding no doubt in part modeled *Westward Ho!* and his 1831 play *The Lion of the West* on the first of Cooper's Leather-Stocking Tales, but it was Paulding, who in his popular 1818 poem, *The Backwoodsman* challenged American authors to write on American subjects. The lengthy narrative poem begins with the lines: "My humble theme is of a hardy swain,/ The Lowliest of the lowly rural train,/ Who left his native fields afar to roam,/ In western wilds, in search of happier home," 7. Cooper so admired the poem that he chose some lines from it to serve as an epigraph to *The Pioneers*. Paulding and Cooper were also members of the same New York City club for the cultural elite, the so-called "Bread and Cheese Lunch"; see Franklin, *James Fenimore Cooper: The Early Years*, 368.

9 Williams, "Deleuze on J.M.W. Turner: Catastrophism in Philosophy," 233–4.

10 Williams, "Deleuze on J.M.W. Turner: Catastrophism in Philosophy," 233.

view invites us to rethink the placement and function of a calamity as something other than an end point or termination of a locale's evolving identity. In this view, disasters, even those as devastating as the earthquakes that hit the New Madrid area, do not destroy a location's identity as much as serve as a stage in a larger process of a dynamic and ever-fluctuating environmental distinctiveness. The lesson is in part that disasters inevitably initiate new stages of regional identity, which continues to change as the territory is subjected to later disturbances. Focusing on the treatment of the New Madrid Earthquakes in Paulding's *Westward Ho!* helps explain why the literary references to the earthquakes were so limited and so ineffectual in furthering the cultural placement of the New Madrid area within an increasingly diverse portrait of the early republic South.

The chapter in question begins when the steamboat carrying the book's hero, Dudley Rainsford (who is heading to New Orleans to purchase furnishings for his new home in the Kentucky wilderness), makes a stop in New Madrid to replenish its stock of wood. Paulding describes the obscure village and its surroundings in a manner that emphasizes its status as a place without much to recommend it and almost equally devoid of a regional social identity. The area is something of a non-place, a wild land that is not quite a wilderness, a settlement that is not quite a town. And over the scene hangs literally an atmospheric pall, a veil of heat and vague foreboding that anticipates the cataclysm to come. As Paulding writes, "It was a close sultry day, with scarcely a breath of air stirring, and the atmosphere of a hazy obscurity, which almost always lays a load of languor on the spirits."[11] Beyond serving as a descriptor for the climate, this so called "hazy obscurity" permeates other aspects of the scene, even affecting the animals: "The birds were sheltered in the deep forests, where they remained panting in silence; and the few domestic animals to be seen, ventured as far into the rapid stream as they dared, and there stood lashing the insects with their tails, listlessly and languidly, as if the effort was almost beyond their strength" (126). Such descriptions result in a narrative set in an unstable and ill-defined location that seems not only to be waiting for and even anticipating the event to come, but somehow reliant upon that event to pull the scene out of its "hazy obscurity" and into a more solid identity.

Soon enough this pervasive languor is broken by the earthquake, an event that instantly invites us to begin to *place* the area, both in time and space. But the narrative instability persists, for as soon as we try dating the scene based on what we know historically as the first of the New Madrid earthquakes, we realize that Paulding is way off. He describes the day as "a close sultry day, with scarcely a breath of air stirring," a day in which the livestock take shelter from the insects by dipping themselves in the river, but we know that the first of the several dramatic quakes was on December 11, 1811, in the dead of winter. Thus even from our privileged historical position, our understanding of the event in time and place remains unsettled because Paulding locates the quake at the opposite end of its historically accurate moment.

---

[11] Paulding, *Westward Ho!*, 126. Subsequent references appear parenthetically in the text.

Such a discrepancy between fact and fiction contributes to what appears to be a larger project of radical displacement that follows. Once ashore, Rainsford and Captain Sam stand and admire the "Great Prairie ... which extends for many miles from the borders of the Mississippi" (126). As they do, they hear a "rumbling" and immediately after

> the earth opened between them, and they stood rocking to and fro on either side a yawning chasm. The ground rose in waves, like the sea in a storm; the vast trees that skirted the bare precincts of the endless plain nodded and struck their high heads together with a crash, and lashed each other with their giant limbs; the earth burst its strong ribs, and rose, and split into vast ravines; the waters broke through their bounds, and while they formed new lakes, or forced themselves into new channels in some places, in others they left large spaces high and dry. (127)

While Paulding has the season of the earthquake wrong, his description of the catastrophe mirrors that of those who experienced the event first hand. In fact, much of his account derives from Timothy Flint's 1826 travelogue, *Recollections of the Last Ten Years*, a book Paulding lauds in his preface and from which he extracts source material for much of the novel.

Essentially the earthquake results in the wholesale displacement and replacement of the New Madrid area transforming plains into ravines and dry ground into lakes. Notably, in describing the terrible violence of the quake, Paulding takes pains to demonstrate how the disaster instantly violates the way we think about the land: shifting it from solid to liquid as the ground swells and breaks in waves "like the sea in a storm." The upshot of this description is that while Paulding's quake is the most dramatic event in his book, it seems designed to further dislocate the New Madrid area by not only occurring at the wrong time of year but also by literally reshaping the land as the narrator strives to orient or place the area within the tale. Paulding emphasizes this sense of dislocation in his portrayal of the quake's aftermath in which, "Darkness, or at least an obscurity, like that of a total eclipse of the sun, came over the world" (128). Such a darkness did actually descend upon New Madrid after the initial quake, and Paulding makes much of the event, even addressing how the darkness confused "all animated nature" (128). He then turns from his portrait of a wilderness in chaos to the town of New Madrid itself, and this brief mention amounts to the only glimpse Paulding offers of the settlement. In surveying the horrid scene of devastation, Rainsford sees "houses tottering and tumbling to pieces, and the people fleeing to and fro in all the desperation of overwhelming terror" (128).

Of all of the disturbing events Paulding conjures up in his portrayal of the quake, the most effective in creating a sense of how the catastrophe is both psychologically and geologically unsettling occurs when he describes the quake's effect on the New Madrid graveyard. He writes that the quake, "spared neither the living nor the dead, for all at once he saw the little graveyard of the village, with its mouldering bones and quiet inhabitants, lifted, as it were, from its resting-place, and hurled into the torrent, where it and they were scattered, never to be associated again in time or in eternity" (128). Here and elsewhere in his brief mention of the disaster, Paulding reinforces the idea that the quake has not only shattered the physical landscape, it has also shocked the community's psychic and even spiritual realm. It is not enough that their humble village has been leveled; the quake must take a toll even upon the dead by casting the bones of New Madrid's graveyard forever into oblivion. The total effect of the quake upon the living and the dead, the human and the non-human is of dislocation and disassociation.

Paulding takes from Flint much of his description of the earthquake and its effects upon the landscape, even the destruction of the New Madrid graveyard. But what is of perhaps greater interest is what Paulding chooses to omit from Timothy Flint's account. Notably he excludes almost all references to the actual population of New Madrid, leaving out Flint's description of the human response and the makeup of the New Madrid population, which reads as follows:

> One result of these terrific phenomena was very obvious. The people of this village had been noted for their profligacy and impiety. In the midst of these scenes of terror, all, Catholics and Protestants, praying and profane, became one religion, and partook of one feeling. Two hundred people, speaking English, French, and Spanish, crowded together, their visages pale, the mothers embracing their children.[12]

Flint goes on to relate several individual accounts of the earthquake, including one from a "French gentleman."[13] While the Reverend Flint's reading of the earthquake as divine punishment for a wayward people is perhaps to be expected from a minister, Paulding seems equally uninterested or unwilling to relate other factors associated with the human side of the catastrophe. He, for the most part, imagines the destruction of the quake in accurate terms, but he seems unwilling to allow us a closer look at the actual people of the town or of their personal reaction to the event. For example, he makes no mention of New Madrid's multiethnic population. In almost every instance in this short episode he severely limits his treatment of the human toll to what Rainsford views from a distance. This technique contrasts with Flint's account and with Paulding's much more detailed descriptions of the novel's other locales, making it hard not to read his quake as commentary on how such a dramatic event actively displaces the modest river settlement on both the natural and human scale.[14]

It also demonstrates how *Westward Ho!* relates to the thematic concerns of Paulding's famous 1818 narrative poem, *The Backwoodsman*. Derrick Spradlin convincingly argues that Paulding's poem reveals "how the literary figure of the frontier traveler was used as a means to comment on the possibilities and strategies

---

[12] Flint, *Recollections of the Last Ten Years*, 224.

[13] Flint, *Recollections of the Last Ten Years*, 225.

[14] Richard Drinnon makes a strong case for *Westward Ho!* as an Indian-hating novel, and the instances of racial violence and general racist portrayals against not only American Indians but also African Americans opens the possibility that Paulding omits mentioning the diversity of New Madrid's citizenry out of a desire to whitewash the population; see Drinnon, *Facing West: The Metaphysics of Indian-Hating and Empire-Building*, 119–30.

for the survival and development of the nation," meaning that texts like Paulding's *The Backwoodsman* employ "the frontier as a distant arena where nationalist potentialities are put to the test; they use the figure of the traveler as a vehicle for U.S. culture on the frontier, the vast and wild unknown that lies beyond the bounds of social hegemony."[15] While this statement fits the general narrative direction of the Dangerfield family in *Westward Ho!*, it does not seem to hold true for Rainsford's frontier experience of the New Madrid earthquake. The event for the most part effaces both personal and national identity, erasing lives and historical memory alike (as symbolized by the destruction of the cemetery). Paulding's earthquake serves then as a reminder of the power of the natural world to, under extraordinary circumstances, shake off the trappings of human settlement and with it render meaningless human sentimentality and nationalistic and regional distinctions. Paulding's reimagining of this dramatic quake invokes, unsettles, and then further obscures the cultural definition of this particular corner of the broad canvas of early America.

Surprisingly the quake does not—at least in Paulding's tale—seem to have any impact on the world beyond the immediate area, and he emphasizes this situation when he has Rainsford—who turns around and returns home to Kentucky—relate his experience to his friends only to learn that they have not even heard of it. Hence Paulding's dramatic portrayal of the earthquake starkly contrasts with the effects of that cataclysm on the world beyond the immediate territory, a situation that promotes the Deleuzian idea that while such seemingly apocalyptic catastrophes can dramatically recast the identity of the local territory, they do not seem to bring about meaningful change to the realm beyond the epicenter of the disaster. In the case of Paulding's novel, by selecting the New Madrid earthquake as the event in question, he underscores how little impact even a cataclysmic event has on the larger world.

Cole's Disaster Portraits

At the same time that writers such as Paulding, Cooper, and others were penning historical romances presenting sublime environmental scenes infused with and impacted by natural disturbances, the renowned British-American painter, Thomas Cole was doing much the same on his canvases. While Cole, himself influenced by the disaster-laden paintings of J.M.W. Turner,[16] does not seem to share James

---

[15] Spradlin, "Westward Expansion, Indian Subjugation, and the Frontier Gazes of Jackson Johonnet, Zebulon Montgomery Pike, and James Kirke Paulding," 39.

[16] For mentions of Thomas Cole's debt to the work of J.M.W. Turner see Elizabeth K. Allen's *Open-Air Sketching*, 32; Barbara Babcock Millhouse's *American Wilderness*, 28, and Elliot S. Vesell's edition of Louis Legrand Noble's *Life and Works of Thomas Cole*, which contains a March 1, 1830, letter Cole wrote to his patron Robert Gilmor in which he states, "I should consider Turner as one of the greatest Landscape painters that ever lived," 85–6.

Williams's interpretation of the Deleuzian "world-view where catastrophe plays a positive role," he does, however, adhere to the idea that "no well-defined territory is safe or static."[17] As an examination of Cole's best known landscape paintings of the 1830s suggests, his work repeatedly embraces the idea that disaster signifies a natural order and power associated with the land and that humanity's development of the land itself likewise serves as a damaging engine of change. In Cole's best known paintings we are presented with an environmental lesson, one in which the destructive tendency of mankind is compared with nature's own sublime ferocity.

Cole counted Cooper as one of his friends and mutual influences, and the mark of each man's impact on the other is fairly easy to see. In 1827, after reading *The Last of the Mohicans*, Cole was struck with Cooper's wilderness fantasy realm, and he painted two scenes from that book: *Scene from The Last of the Mohicans* and *Cora Kneeling at the Feet of Tamenund*. Cooper likewise was much taken with Cole's work, and it would seem he was especially fascinated with Cole's five-painting masterpiece *The Course of Empire* (1834–1836), which—influenced by Turner's painting *The Decline of the Carthaginian Empire* (1817)—depicted on a local scale the birth, rise, fall, and ruins of an unspecified human civilization against a landscape that changes in tandem with the human domination of the environment.[18] The romantic painting imparts a realistic cautionary tale, one that readily melded with Cooper's larger project of creating an allegorical past meant to comment on what he saw as the new nation's troubled present and uncertain future. This influence emerges in much of Cooper's work, especially in books such as *The Prairie* (1827), *The Deerslayer* (1841) and *The Crater* (1847) that all deal with the problem of reconciling the community-centered national imperative of Manifest Destiny with its companion myth of the American self-made, independent, white frontiersman who (as embodied in Natty Bumppo) simultaneously enables and abhors the settlements that follow from his pioneer ways.

In the same year that Cole completed *The Course of Empire*, he also finished a stand-alone piece titled, *View from Mount Holyoke, Northampton, Massachusetts, after a Thunderstorm*, a painting best known as *The Oxbow* (Figure 6.1). Portraying the Connecticut River Valley from the top of Mount Holyoke, this painting, perhaps his most famous work, depicts in stark contrast the way that human settlement of the valley below has drastically altered the landscape.

The painting is divided into two halves that form the edge of a conflict between the wilderness on the left and the human-tamed valley on the right. The right side is characterized by a pastoral scene of farms, houses, roads, and a massive project of deforestation, with further cutting of trees evident in clear-cut areas that scar the distant mountainsides. The left-most section of the painting depicts a wild scene where the thunderstorm of the painting's title moves over a still wild landscape of blasted trees and rocky outcroppings toward the settlement in the valley. At the center of the painting, all but hidden from view, is an artist at work at

17 Williams, "Deleuze on J. M. W. Turner: Catastrophism in Philosophy," 233–4.

18 Hell, "Imperial Ruin Gazers, or Why did Scipio Weep?," 175–6.

Fig. 6.1 Thomas Cole (American, 1801–1848), *View from Mount Holyoke, Northampton, Massachusetts, after a Thunderstorm—The Oxbow*. 1836. Oil on canvas, 51½ × 76 inches. The Metropolitan Museum of Art. Gift of Mrs. Russell Sage, 1908. (08.228). Image © The Metropolitan Museum of Art, New York, NY, USA.

an easel facing the direction of the viewer. Rochelle Johnson—who makes much of Cole's impact on the nature writing of James Fenimore Cooper's daughter Susan—persuasively argues that the division portrayed in this now-famous (but during its day lesser known) painting demonstrates "Cole's sense of imbalance" and the recognition of a "wilderness condition that must give way to accommodate the idyllic rural scene. That idyllic scene comes at a price: the destruction of wilderness."[19] Pointing to Cole's gesture of placing an artist in the midst of the painting itself, Johnson writes, "While Cole's depiction ... seems a fairly mimetic representation, the presence of an artist within that landscape suggests that the very landscape he represents is to some degree deceptive—that it is a result of craft."[20] Examined more fully, however, it would seem that by placing the image of himself (or at least that of an artist like him) in the painting Cole reminds the viewer that such landscapes are themselves portraits of nature and man's influence upon the land. After all, the placement of a landscape painter in a painting of a beautiful landscape might be said to actually enhance the realism of its portrayal.

By positioning the artist in the wilderness but at the center of the painting communicates at least three ideas key to its presentation of the environment. The first is in keeping with Johnson's argument that the painter is not a bit-player here but is instead the figure really at the center of the arrangement. Secondly, by placing the image of a painter at the center of the painting Cole reminds us that the painter, whatever his affinity for the wilderness, does not live there, and ventures into its hostile environs only as long as it takes to adequately capture the distinction between the transformed landscape below and the soon-to-be transformed wilderness. In this respect the artist in the painting takes on a role similar to that of the people in Cole's *The Course of Empire* who begin humbly in the so-called *Savage State* and then due to their success in realizing an alternative landscape (one ostensibly better for human needs) so change the land as to make it nearly unrecognizable before they destroy themselves. The artist then, as argued by the placement of just such a man in *The Oxbow*, becomes something of a different kind of frontiersman or pioneer. Unlike Cooper's Natty Bumppo, Cole's painter may not chose to live in the wilderness (indeed he is about to be drenched in a powerful storm that looks more than a match for his umbrella), but in imagining it, portraying it, and thereby commodifying it, the artist becomes complicit in its destruction or transformation from a non-human wilderness to a human-centered environment. And thirdly, Cole, by directing his artist's tiny, paint-speck face toward the viewer and not the sublime scene behind him, suggests that the painter is looking at us from beyond the frame, a mirror of humanity in a scene marked by the pretty pastoral desolation of humanity's making. The painter is looking at us, and therefore becomes a proxy for the scene he has painted and that so dramatically reflects the human urge to create wholesale environmental disturbance. Cole carefully details the way such disturbance looks from afar, a

[19] Johnson, *Passions for Nature*, 87.

[20] Johnson, *Passions for Nature*, 88.

point of view shared by the wilderness and the approaching storm itself. We see the unnaturally straight lines of farms and roads cut into the landscape below; we see the cryptic script of clear-cuts on the mountainsides in the distance; we see the general denuding the forest throughout the scene; and we see the thin trails of chimney smoke from the houses. Then to the left of the painting we are reminded of what came before. Like *The Course of Empire*, Cole's *The Oxbow* gives us a before-and-after picture of the unsettling nature of human settlement, and it does so in a condensed fashion, in a single painting, that makes the artist himself figure for all of humanity. "Cole's message," writes John Opie,

> was that civilization was a futile revolt against nature. He ambitiously endeavored to combine nature, history, and national identity in his art ... The objective of the landscape art of Cole and his contemporaries was to bring Americans back to nature as the central source of a personal and national identity. Nature is permanent, whereas human affairs simply swing through constant historical cycles.[21]

Aaron Sachs has also recently pointed to the theme of destruction that resides at the core of several of Cole's deceptively pastoral scenes:

> Thomas Cole, one of America's consummate Arcadians, not only reminded his compatriots that "The Course of Empire" usually ended in "Desolation," but also invoked the legend of the Flood to re-pose the question of how we might rebuild a good society after bringing cataclysm down on our heads yet again.[22]

As Sachs notes, this flood imagery is found not only in *Desolation*, but also in Cole's *Subsiding of the Waters of the Deluge* and *The Oxbow*.[23] Conceding its message of environmental anxiety, Carolyn Merchant does, however, read *The Oxbow* in a more sophisticated and hopeful way. She writes,

> The logging scars in the forest on the background hill appear to be the Hebrew letters for Noah. When viewed upside down from a God's eye view, these letters form the word *shaddai*, or 'the Almighty.' God's presence in the landscape underscores his covenant with Noah and anticipates the final reunion of God and the earth at the parousia. This painting suggests that humans can redeem the land itself as garden, even as they redeem themselves through laboring in the earth. They can reinvent the original Eden by transforming wilderness into a pastoral garden. But the scars on the hillside herald the decline of the garden and the smokey haze rising from the forest the end of unblemished wilderness.[24]

---

21 Opie, *Nature's Nation*, 203–4.

22 Sachs, *Arcadian America*, 344.

23 Sachs, *Arcadian America*, 344–5.

24 Merchant, *Reinventing Eden*, 93–4.

We learn more about Cole's theory of landscape art and the human role within that aesthetic from his "Essay on American Scenery," which was published in *The American Monthly Magazine* in January 1836. Cole writes that while some may argue that the American landscape is less beautiful than those found in Europe because it "possesses little that is interesting or truly beautiful—that it is rude without picturesqueness, and monotonous without sublimity—that being destitute of those vestiges of antiquity whose associations so strongly affect the mind, it may not be compared with European scenery."[25] Cole counters that idea, writing that it is the general lack of such "vestiges of antiquity," that is, ruins of ancient human habitation, that give it its power. He writes, "the most distinctive, and perhaps the most impressive, characteristic of American scenery is its wildness."[26]

Cole imparts just such an American wildness in his 1831 painting, *Tornado in an American Forest*, which presents a storm-wracked forest scene where chaotic winds are destroying the trees, and where are a solitary man clings for safety to the blasted trunk of a tree (Figure 6.2). As this painting proves, *The Oxbow* was not the first work in which Cole placed a single human figure within a stormy wilderness. However unlike *The Oxbow*, this earlier painting portrays an entirely non-human environment devoid of any evidence of roads, farms, or houses; the figure of the man himself is the painting's only sign of humanity. Interestingly, like *The Oxbow* the environment has still undergone a form of deforestation—albeit a natural form—from what appears to be the shearing of the trees by the tornado.

The ruined trees with their top branches ripped off serve as the focal point of the painting and the man holds fast to one of these trees, his clothes and hair pulled back by the wind. The figure holds on for life in a wilderness made even more treacherous by the devastation of a tornado. It is a portrait of a man within a wilderness within another wilderness. The human place in this realm is tenuous and his survival is uncertain. Whereas *The Course of Empire* and *The Oxbow* are more allegorical paintings that present environmental cautionary tales about the destruction that follows human development of the land, *Tornado in an American Forest* lacks any such overt lesson.

Cole is far more forthcoming about his environmental opinions in his essay, especially at its conclusion where, after spending several pages favorably comparing the beauty of American mountain, river, and forest scenes with those of Europe, he ends with a lament that might serve as the thesis for several of his paintings of that period:

> I cannot but express my sorrow that the beauty of such landscapes are quickly passing away—the ravages of the axe are daily increasing—the most noble scenes are made desolate, and oftentimes with a wantonness and barbarism scarcely credible in a civilized nation. The way-side is becoming shadeless, and another generation will behold spots, now rife with beauty, desecrated by what is called improvement; which, as yet, generally destroys Nature's beauty

---

25 Cole, "Essay on American Scenery," 4.

26 Cole, "Essay on American Scenery," 5.

Fig. 6.2 Thomas Cole (American, 1801–1848), *Tornado in an American Forest*. 1831. Oil on canvas. 46 3/8 × 64 5/8 inches. Corcoran Gallery of Art, Washington, DC. Museum Purchase, Gallery Fund, 77.12.

> without substituting that of Art. This is a regret rather than a complaint; such is the road society has to travel; it may lead to refinement in the end, but the traveller who sees the place of rest close at hand, dislikes the road that has so many unnecessary windings.[27]

Cole presents the human development of the American landscape as the true catastrophe, as an aesthetic disaster that leaves only ugliness in its path. Rochelle Johnson argues, "Cole's works therefore contain a paradox that remains at the heart of American identity: the nation loves its wild lands, but it loves them to death, as it were."[28] While this paradox does not easily extend to a painting like *Tornado in an American Forest* it easily fits *The Oxbow* and *Course of Empire*, offering a way of seeing the irony inherent in mankind's pastoral impulse. This same paradox also appears in several of the written texts of the time, and one need not look too long to see it in Cooper's novels. Like Cooper in *The Pioneers*, Cole simultaneously mourns the loss of the wildness of the landscape while also conceding that such destruction is inevitable in a nation bent on economic development. He is willing to believe that such spoiling of the wilderness is "the road society has to travel," but he refuses to applaud the trend. Like so many people of his day, and even of ours, Cole saw the ruin of the natural world as the price America had to pay for its success, but as paintings like *The Oxbow* and *The Course of Empire* series argue, by travelling that road of environmental destruction, humanity risks much more than an ugly landscape. It risks its own existence.

Environmental Morality and *The Crater*

Upon viewing Cole's *The Course of Empire*, it would seem Cooper picked up on the artist's environmental warning, and in 1847 he published *The Crater*, a novel that explored the problem of colonization from a literary angle. When viewed from our twenty-first-century vantage point it is easy to forget that Fenimore Cooper was one of the most experimental writers of the nineteenth century. By the time he published *The Crater* in 1847 he could look back on a long career marked by literary innovation, great success, and, as is often the case with artists who strive for originality, a good deal of literary and economic failure. His early successes, such as *The Spy* (1821), *The Pioneers* (1823), and *The Pilot* (1823) defined, or at least in some areas forced a redefinition of, American literary prose. In just those three books, written in as many years, he created the templates for the American historical romance, the frontier romance, and the maritime novel, templates that would not only benefit his immediate predecessors such as Nathanial Hawthorne and Herman Melville, but that also provide the foundation for today's political thrillers, westerns, and sea tales.

---

[27] Cole, "Essay on American Scenery," 12.

[28] Johnson, *Passions for Nature*, 77.

One of the effects of having made such an early impact on his nation's literature was that Cooper lived long enough to refine his own inventions in light of new discoveries, and to also witness how those writers who followed in his footsteps approached the genres he did so much to create. As Hugh MacDougall writes, in composing *The Crater*, Cooper "relied heavily on the records, published in 1845 of the first American scientific expedition to the Pacific Ocean, led by his US Navy friend Charles Wilkes, and was encouraged by the success of Herman Melville's first Pacific novel, *Typee* (1846)."[29] For Cooper, the Pacific was a new maritime literary frontier, a relatively unexplored realm, both in reality and literarily, in which he could set the action of *The Crater* a book that is, like much of Cooper's work, as much an intellectual discourse on sociopolitical ideas as it is a thrilling tale of oceangoing adventure. That Cooper had sought out the new waters of the Pacific for his narrative while at the same time exploring what was for him well-trodden intellectual ground is evidenced by his creation of a story that, as he had done in *The Prairie* (1827) and his three Littlepage novels (1845–1846), fully engaged the problem of human cultural succession. "The result," writes MacDougall, "was a story that is partly vivid shipwreck adventure in the Pacific, partly the presentation of a utopian community, and partly an impassioned criticism of American society—much of it still relevant today."[30] So it was, that late in his career, Cooper returned to the sea tale to further advance it from its earlier romanticized portrayals to a vehicle for his continued discussion of substantial sociopolitical ideas, and nowhere in his writings had Cooper so thoroughly and productively incorporated the concept of the natural disaster into the problem.

To briefly summarize the book, *The Crater* tells the story of Mark Woolston, a young Pennsylvania lad who in 1793 joins the crew of the merchant ship the *Rancocus*. The first of several of the novel's disasters strikes when the ship encounters a storm as it enters some uncharted Pacific reefs. The morning after, Mark and another crewman, Bob Betts, find themselves the only remaining people on the ship, which has become inextricably trapped among some barren reefs. Shipwrecked, the men choose to make a new home in what is actually the cone of a dormant volcano, and using provisions and equipment from the ship begin to cultivate the floor of the old crater. With help from the livestock and agricultural supplies on the ship, their efforts pay off, and they soon have plenty of food. But then a hurricane sweeps Bob out to sea in one of their smaller boats, and a third major disaster follows when an earthquake and the eruption of a volcano out at sea, raises the surface of the island several feet and creates a one-thousand foot high volcanic peak that Mark names Vulcan's Peak. Eventually, Mark and Bob are reunited, and, with the help of a new ship, make trips back home where they get supplies and enlist volunteers to create a permanent settlement at the crater island. As their settlement matures and prospers, Mark (now Governor Woolston) and his group must battle hostile natives, a pirate attack, and finally an increasing amount

---

[29] MacDougall, *The Cooper Bookshelf*, n.p.

[30] MacDougall, *The Cooper Bookshelf*, n.p.

of jealous infighting amongst the settlers. Frustrated by the infighting, Mark and his allies leave for an extended visit to the United States. Upon their return they find no trace of the crater island or its colonists. Searching the area they finally locate the tip of Vulcan's peak jutting through the water. They then realize that the worst of the book's disasters occurred in their absence. An intense earthquake has drowned the entire colony. The settlement and the bones of its 500 settlers rest several fathoms below the water's surface.

From its beginning, *The Crater* presents us with a complicated view of how natural disasters are coupled with, and often exacerbated by, a host of humanity's failings. In the first such case, rain and darkness impede the vision of the men aboard *The Rancocus*, and thus cause the vessel to founder on some uncharted breakers while also forcing the loss of all but two of the sailors in the process. While their impaired vision is the most directly physical part of the problem, Captain Crutchley is the figure most deserving of blame because, as Cooper explains, it is his habit to "drink rather too much grog, at his dinner,"[31] and, despite the vehement warnings of the ship's first officer, Mark Woolston, the captain errs by refusing to wait until daylight before progressing. Terrified by the captain's recklessness, Bob Bett's, the ship's best lookout, (who is initially reluctant to report his misgivings) now pleads with the captain that "It's my opinion, sir, that the ship is in the *midst* of breakers at this instant, and that we may go on 'em at any moment!" (42). Crutchley refuses to listen, exclaiming, "The devil it is!" in the very instant the ship runs upon the breakers (42). Cooper writes,

> Just as the ship had got dead before the wind, and was flying away to leeward, short as was the sail she was under, the atmosphere seemed to be suddenly filled with a strange light, the sea became white all around them, and a roar of tumbling waters arose, that resembled the sound of a small cataract. The ship was evidently in the midst of breakers, and the next moment she struck!" (42)

Doomed by the mistakes of their inebriated captain—who is perhaps justly swept overboard in the ensuing disaster—the ship then plows into the uncharted breakers hidden in the darkness ahead of them. This shipwreck, which places Mark and Bob on the volcanic island that serves as the setting and controlling metaphor of the book, is a small example of the much more dramatic catastrophe to come. As Cooper's presentation of the shipwreck's causality demonstrates, it is not simply the result of poor visibility and bad luck; it is a disaster born of human error as well. The captain's judgment is hampered by drink; their otherwise trusty charts are in this sole respect inaccurate; and Bob Betts, who, along with Mark, does recognize the danger ahead, is at first too afraid of displeasing the captain to accurately report his misgivings. In the morning following the storm Mark and Bob realize they are the only ones who were not swept out to sea. They find a spot of land, a desolate reef, in the distance and laboriously steer the *Rancocus* in that direction.

---

[31] Cooper, *The Crater*, 34. Subsequent references appear parenthetically in the text.

With the first major natural disaster of the novel over, Cooper then weaves within the narrative a series of intriguing links between the problem of settling the land, the threat of further disasters, and the humane treatment of animals. For anyone who comes to the book expecting a simple tale of shipwreck adventure, *The Crater* makes for a pleasantly surprising read in that it melds an intriguing adventure narrative with an increasingly pronounced work of social and environmental commentary. The narrative becomes a compelling thought experiment, with the island, by its very geology, taking on the role of a form of beaker or petri dish for this mixture of desperately inventive humans stranded in a seemingly sterile environment. This experiment begins when the men discover at the center of the barren volcanic island a pronounced mound. They ascend the steep side of the mound with difficulty and find, in Cooper's words, that "Here a surprise was ready for them, that drew an exclamation from each, the instant the sight broke upon him. Instead of finding an elevated bit of table-rock, as had been expected, a circular cavity existed within, that Mark at once recognized to be the extinct crater of a volcano!" (60). Upon further inspection, the men realize that the singular crater is entirely surrounded by the wall of volcanic rock at an even height of about 60 feet, and that the interior of the crater contains something in the range of a hundred acres of walled in space. In a single spot at the base of the wall they find a small, natural hole, a passage that Cooper describes as "a sort of gateway, by which to enter and quit the cavity" (60). The empty bowl of the crater also immediately conveys a sense of foreboding, at least in terms of its relationship with the living world, when Cooper ends the chapter with Mark's disconcerting observation that

> from some cause or other, the birds avoided the crater. It really seemed to him that their instincts warned them of the dangers that had once environed the place, and that, to use the language of sailors, "they gave it a wide berth," in consequence. Whatever may have been the cause, such was the fact; few even flying over it, though they were to be seen in hundreds, in the air all round it. (61)

That the danger that seems to keep the birds away might not have entirely passed is left unsaid by the two men. Regardless, with their ship essentially stuck in the reef around the island, and their inability to sail the large ship on their own for any great distance, they decide that their best chance for survival will be to, as Bob puts it, "Robinson Crusoe it" on the reef and hope for rescue (63). Resolved to make a go of it there, the two young men guide their still-intact but trapped ship closer to the shore and with it the ship's store of enough beef, pork, bread, and flour to last the men for an estimated "five or six years" (73).

A key part of Cooper's experiment with *The Crater* is to provide the men with the supplies necessary for the cultivation, and indeed the terraforming, of the barren reef island and its odd crater. So while they have enough food, they also realize that to avoid scurvy they need to cultivate fresh vegetables. Thus Cooper gives the men a good reason to engage in the hard work of attempting to farm what is essentially barren rock by creating a need that only the land can provide. Although the rains will supply adequate water and the ship has plenty of preserved rations, the men know that their survival is still dependent upon their ability to grow something in this challenging environment. In the fashion of most nineteenth-century romances, Cooper allows himself the liberty to give his protagonists a convenient assemblage of items that make their colonization of the land possible. As it turns out, their Quaker-financed ship was originally stocked with the intention that it serve as a first step in a combined humanitarian and trading mission to the natives of several islands, it was laden with gardening implements, seeds of all kinds, and some livestock as well. As Cooper writes, Captain Crutchely brought along these agricultural necessities "as a benefit conferred on the natives of the islands he intended to visit, and through them on future navigators. This care proceeded from his owners, who were what is called 'Friends,' and who somewhat oddly blended benevolence with the practices of worldly gain" (78). The ship's modest collection of farm animals, including chickens, ducks, four shoats, and a goat named Kitty was the product of the captain's preference of having fresh meat and a regular supply of goat's milk with his tea.

Mark and Bob, despite their own problems, worry over the comfort and humane treatment of the animals, and their concern for the creatures goes well beyond the mere fact that in taking care of the creatures those animals will be more likely to benefit the men. Indeed some of their decisions about how to better care for the animals (such as letting the goat run wild) seem to work counter to the men's best interests because it means they will have to chase down, or even risk losing control over, some of the creatures altogether. The reason that the castaways care so much about the ethical care of their nonhuman charges appears to stem from a largely newfound empathy that springs from the men's own dire situation.

Through Mark and Bob's treatment of the animals, Cooper forwards the idea that when suffering the aftermath of a disaster, survivors (or at least those of admirable character) become more aware of the need to treat animals with care. One of the first and more pronounced mentions of animal treatment occurs when Bob suggests "the necessity of getting food and water ashore for the pigs, as an act that humanity imperiously demanded of them; not humanity in the sense of feeling for our kind, but in the sense in which we all ought to feel for animal suffering, whether endured by man or beast" (80). Here Cooper enriches the idea of humanity to encompass a responsibility to both "man and beast," even when that responsibility means seemingly unnecessary expense in time, energy, and resources for the human caretaker. That Cooper classes "man" within the category of animals is particularly notable, as it somewhat levels the hierarchy of being and acknowledges how both groups are sometimes cursed with "animal suffering." We find another notable instance of such sentiments a little later, when Mark watches the newly released ducks cavorting in the pools of water on the island. He watches them "enjoying themselves" and "a doubt first suggested itself to his mind concerning the propriety of men's doing anything that ran counter to their instincts, with any of the creatures of God" (83). This epiphany about the kind treatment of birds is a complicated one, in which Mark admits to himself that he had always been bothered by seeing birds in cages, especially because "[t]hey were created with an instinct for flight, and intense must be their longings to indulge in the power which nature had bestowed on them" (83). In considering his feelings about the ducks, Mark's self-awareness is keen, and he realizes that as a prisoner on the barren island he is now in a position similar to that of an animal in captivity: "In the cage in which he now found himself, though he could run, walk, leap, swim, or do aught that nature designed him to do, in the way of mere animal exploits, young Mark felt how bitter were the privations he was condemned to suffer" (83).

Mark is at his most empathetic when considering the fate of the goat, Kitty, in whose situation he sees a partial reflection of his own. In bestowing upon Mark a particular empathy for Kitty, Cooper invites a more focused examination of how their shared status results in a mirroring of their suffering. As Mark helps unload the contents of the ship onto the island, he takes a moment to consider Kitty's fate. He knows that he might as well just slaughter her; the goat is after all without a mate and poses a sure threat to the well-being of their hard-earned garden. Yet, Mark finds himself succumbing to humanitarian reservations similar to those Bob invoked in the case of the pigs:

> Nothing with life now remained in the ship but the goat, and Mark thought it best not to turn her ashore until they had greater facilities for getting the necessary food to her than the dingui afforded. As she was not likely to breed, there was no great use in keeping this animal at all, to say nothing of the means of feeding her, for any length of time; but Mark was unwilling to take her life, since Providence had brought them all to that place in company. (82)

Kitty is more than just an object to be milked or slaughtered as needed for their human survival, she has become something of a companion, a pet of sorts, but one whose value exceeds mere amusement. As Mark points out, as the sole goat on the island, Kitty is "not likely to breed," a situation shared by the two men and of particular note to the newlywed Mark, whose separation from his bride, Bridget, vexes him. Even for those animals with mates, the Crater, at first, is no Eden, and the men and animals find commonality in their resolve to survive on an island that seems antithetical to the fostering of any kind of life. This situation changes later in the novel when Bridget and others join Mark and Bob on the then expanded island, and Bridget—repeatedly referred to as an Eve figure—even names part of the island "Eden" (206). In the early stages of the novel, however, Mark finds a certain pleasure in anything that is alive, even the almost useless Kitty, whom he finally decides to spare at least for the moment because "she might be a pretty object leaping about the cliffs of the crater, giving the island a more lively and inhabited appearance, though he foresaw she might prove very destructive to his plantations, did his vegetables grow" (82–83).

Despite his phrasing, Kitty has ceased to become a mere "object" for the men; she is a sign of life in a desolate realm, a partner in their shared suffering, and it is this very suffering that functions as the key to the empathy needed to speed

Kitty's transformation in the men's minds from a mere animal/object to a pet/comrade. Jeremy Bentham in his *An Introduction to the Principles of Morals and Legislation* (1789) charges that animals have been historically treated as lesser beings because humans have imposed the wrong criteria for what makes an animal deserving of rights and humane treatment. Making a connection between the dehumanizing of humans that allowed for slavery and the problem of cruelty to animals, Bentham writes, "animals on account of their interests having been neglected by the insensibility of the ancient jurists, stand degraded into the class of *things*."[32] He then interrogates the idea that rationality and the ability to speak are the proper criteria for the bestowing of rights and legal representation upon an animal:

> Is it the faculty of reason, or, perhaps, the faculty of discourse? But a full-grown horse or dog, is beyond comparison a more rational, as well as a more conversible animal, than an infant of a day, or a week, or even a month, old. But suppose the case were otherwise, what would it avail? The question is not, Can they reasons? nor, Can they talk? But, Can they suffer?[33]

Like Mark and Bob, rather than deny the fact that humans are animals Bentham instead forwards the idea that humans are given privileged status because we have the capacity to suffer and that other animals who share that capacity should enjoy the same protections. That *The Crater* appears to promote something similar to Bentham's animal rights sentiments reveals again, how advanced Cooper's ideas were in regard to the elements of the natural world, people included. Of course we must not equate Cooper's 1847 sentiments about animals with all of the ideas promoted by the present-day animal rights movement. While possessed of an uncommon care for the suffering of their animals, Mark and Bob still judiciously slaughter their animals, especially the pigs, to provide food for themselves and to control the island's animal population as needed.

Still, the care that Cooper's characters extend to their nonhuman companions is remarkably progressive and born of their equally remarkable circumstances. As the narrative of *The Crater* progresses, Mark and Bob's respectful-but-utilitarian treatment of the animals becomes mirrored in their use of the island itself. As their gardens prosper and the grass covers the once-bald rock surfaces in an

---

32 Bentham, *An Introduction to the Principles of Morals and Legislation*, 234–5. For another good literary example of what at the time was considered the uncommonly beneficial treatment of an animal, we need look no further than Paulding's *Westward Ho!*, in which we see the otherwise selfish and obtuse Colonel Dangerfield taking pity on his old beloved race horse Barebones (a horse who had lost the colonel what amounted to his personal fortune). Paulding presents Colonel Dangerfield as a moronic but kind fellow by having him give the washed up racehorse to a man he knows is a "philanthropist" and who "promised on his word that he [Barebones] should never be degraded to any useful occupation," 64.

33 Bentham, *An Introduction to the Principles of Morals and Legislation*, 236.

inviting green, the men and their loved ones dub the island realm a new Eden, and while they give much credit to God providence for their fortune it is clear that their agricultural success is the product of a land use philosophy attentive to the judicious and thoughtful redistribution and cultivation of resources. So increasingly colonized by a mixture of their hard work, savvy agricultural strategy, and providence, the island environment changes from a barren land to a paradise. Realizing that when together with their loved ones, they have all they need there, the men eventually elect to stay on the island as an official colony formed of their spouses, Brigit's two amiable slaves, and those others who come to their eventual rescue.

Yet despite all their resilience, all of their resources, and all of their luck, Cooper makes it clear that at its core the novel's narrative is driven by a series of natural disasters of varying types and degrees of severity but that almost always introduce a narrative twist that pushes the story towards a cataclysmic end. Each of these natural calamities force the men to adapt, and as often as not, to develop new tools for success, one of those tools being empathy with the nonhuman world. Without the opening storm, the men would not have been shipwrecked, and without the tempest that followed months later they would not have been inundated with dead fish for Mark to use as excellent fertilizer for his terraforming efforts, and Bob would not have been swept away in the Neshamony to eventual rescue and return with Brigit and company. In the interim between the loss of Bob and his return, an exceptionally violent earthquake raises the island several feet and reveals even more opportunity for colonization in the form of a discovery of a conjoined stretch of land (the home of the volcano Vulcan's Peak) which is rich in indigenous animal and plant life. So virtually each major shift in the narrative is driven by an accompanying disaster, and despite their violence (and often because of their force) these disasters affect—in Deleuzian spirit—a positive change for Mark and company. As violent yet amoral ambassadors of change, these disasters force in Mark and Bob a consciousness of shared suffering that leads them per Bentham to empathize with each other, with the animals, and, in a more abstract way, with the islands themselves.

They likewise believe that their experiences on the island are strong indicators of God's providence, and with this increase in belief comes a decrease in mentions concerning animal welfare. In relating the castaway Mark's frame of mind, Cooper's narrator states, "The more deeply we penetrate into the arcana of nature, the stronger becomes the proofs of design; and a deity thus obviously, tangibly admitted, the more profound will become the reverence for his character and power" (140). For Mark an unknowable God deserves credit for their situation, and as the narrative progresses he becomes increasingly devout while also seemingly less concerned with the well-being of animals. Gone are the mentions of Kitty, who effectively disappears from the story. Gone are the concerns about the suffering of the hogs, who are left to their own devices on the island and slaughtered as necessary. For Mark, these entities become replaced with further devotion to God, which seems to rise in tandem with the earthquake-caused increase in land area.

In describing Mark's exploration of how a recent earthquake has increased the size of the island, Cooper writes, "God was there, also, as he is everywhere. This our secluded man found to be a most precious consolation. Again and again, each day, was he now in the practice of communing in spirit, directly with his Creator" (173). In conquering the land, Mark becomes more concerned with religion, and with giving credit to the Creator rather than to the land, its creatures, or himself.

Cooper effectively ends the book with an allegory about poor resource use. Adhering to the myth of America's colonial origins, the first settlers represent a group of good, hard-working people who only seek a new start in a new land, but who, through their need for resources nevertheless degrade the land upon which they depend, and in so doing also engage in practices that jeopardize their souls. As years go by, Mark finds himself the governor of a colony of over 500 people. The population explosion is the result of the influx of new settlers and the growing of existing families, and with this increase of population comes an increase in resource depletion. Mark and his colony must figure out a way to raise money to supply needs the colonists cannot provide for themselves. Their first economic venture is in the collection and shipping of sandalwood. The venture is highly profitable but it is also unsustainable; with overharvesting, the sale of sandalwood becomes impossible (352–3). Trading in sandalwood has an additional danger. It was also considered blasphemous due to its use by what Cooper describes as "heathenish people" in non-Christian spiritual practices (100). Earlier in the novel, Mark even wonders if the *Rancocus*'s original plan to obtain sandalwood was a contributing factor in their shipwreck. (The ship's pious Quaker owners express their own reservations, but the economic reward overwhelms their religious scruples and they tempt God in their search for a profit).[34] Mark and company do so again near the end of the story, when they turn immediately to the sandalwood trade to fill their coffers. That Mark trades in sandalwood even though it goes against both his religious views and his understanding of good land practices places him in company with two other of Cooper's environmental advocates—Marmaduke Temple and Natty Bumppo. Each of these men, despite their legitimate concern for the land, engage in selfish and wasteful environmental enterprises, and this trope represents what Cooper sees as one of the central problems people must surmount when they champion any form of natural preservation.

Once sandalwood becomes unprofitable, they turn to whaling as a substitute, but eventually that resource, too, becomes scarce. ("The whales had been driven off the original fishing-ground, and the pursuit was no longer as profitable as it had

[34] Cooper more than once notes the sinfulness of the sandalwood trade. In one passage he describes the soon-to-be-dead Captain Crutchely's plans, and notes,

> Sandal-wood was his aim, a branch of commerce, by the way, which ought never to be pursued by any Christian man, or Christian nation, if what we hear of its uses in China be true. There, it is said to be burned as incense before idols, and no higher offence can be committed by any human being than to be principal, or accessary, in any manner or way, to the substitution of any created thing for the ever-living God. (33)

been, three fish having been taken formerly to one now" [447]). As the process of colonization continues, unsavory people begin to make their home on the island. Cooper writes, "Unhappily, among these emigrants were a printer, a lawyer, and no less than four persons who might be termed divines" (430). The arrival of these figures, each of whom in Cooper's thinly veiled editorializing represents a threat to the unity of the colony, marks the replacement of the colony's once congenial atmosphere with one of cliquishness, infighting, newspaper smears, and lawsuits. In the months that follow the colonists divide the island into districts, write up a constitution, and vote for a new governor.

At the end of the tale, Mark, Bob, and their growing families prepare to depart for an extended trip back home to Philadelphia. Though they expect to return to the island, it is clear that these original settlers have lost control to the newer mass of colonists. Before sailing away, Mark, Bridget, and company visit the Edenic "plain" of the island, a special place they fear will soon be spoiled by human expansion: "the passengers all ascended to the plain, to take leave of that earthly paradise. Nature had done so much for this place, that it had been the settled policy of Mark Woolston to suffer its native charms to be marred as little as possible" (451). But with the new regime "a law was passed authorizing a new survey, and a new subdivision of the public lands" into residences (451–2). In considering the future of this part of the island chain, Mark thinks to himself, "As yet, its beauties had not been much marred; though the new irruption menaced them with serious injuries" (452). Here Cooper uses the term *irruption* to refer, not to any actual volcanic event, but instead to describe the destructive tendencies of the shortsighted colonists. Through this terminology Cooper firmly links the colonists' unfounded pride, greed, and even their environmental abuses with the massive earthquake that occurs sometime after Mark and company leave. As we will see, this human "irruption" precedes a natural one.

We return with Mark over a year later to find that an earthquake has pulled the island and its over five-hundred colonists deep below the sea, leaving only the tip of the volcano's peak still above water, standing as a marker over the watery mass grave. It is perhaps the most dramatic and chilling conclusion to any of Cooper's novels, and of the scene he writes, "It might be said to resemble, in this respect, that sublime rock, which is recognized as a part of the 'everlasting hills,' in Cole's series of noble landscapes that is called 'the March of Empire'" (456). Cooper is slightly off on the title of Cole's *The Course of Empire*, but in most other respects in *The Crater*, he conveys the lessons of social responsibility and environmental morality found in Cole's masterwork. While in that series of paintings humanity destroys itself through war, in *The Crater* the destruction is ostensibly of divine origin; it is a jealous God's punishment for a colony that has lost sight of the divine in nature, who failed to acknowledge the spiritual design in the bounty of their environment. Cooper's final paragraph reiterates this lesson:

> Let those who would substitute the voice of the created for that of the Creator ... remember their insignificance, and tremble. They are but mites amid millions of other mites, that the goodness of Providence has produced for its own wise

> ends ... Let that dread Being, then, be never made to act a second part in human affairs, or the rebellious vanity of our race imagine that either numbers, or capacity, or success, or power in arms, is aught more than a short-lived gift of His beneficence, to be resumed when His purposes are accomplished. (459)

Cooper presents the earthquake at the end of *The Crater* as divine retribution for a loss of humility, and notably that humility includes the debt that the people owe God for the environment. This brutal punishment is truly of Old Testament magnitude, Mark rounds up his family and friends and departs just in time to avoid the total destruction of the island and its hundreds of men, women, and children, like so many ungrateful mites.

That Cooper so strongly links the violent end of his terraforming adventure/disaster novel with the concept of a vengeful God demonstrates how *The Crater* straddles older religious causation with concepts very much born of the Enlightenment era. Mark and company employ reason and even scientific principles in turning the rocky island into a new Eden, but—as Cooper's narrator argues—it is the pride and greed of the new settlers that so angers God that he punishes them (children and all) with a horrible death. In *The Culture of Disaster*, Marie-Hélène Huet convincingly argues that the nineteenth-century saw a pronounced shift away from thinking that disasters were of divine origin. She writes, "The Enlightenment project is widely credited with the recognition that natural disasters were not sent by a wrathful God but stemmed from the workings of a violent universe."[35] As these Enlightenment ideals increasingly won out against older superstitions, "the vision of a disaster willed by God was swiftly replaced by one of human-engineered calamity. If volcanoes erupted and the earth shook without divine intent, humans made the damages greater and more grievous to bear."[36] Cooper, it would seem, offers a third option. In *The Crater* he champions Mark's and Bob's ingenuity and agricultural acumen while also promoting the almost Puritanical idea that divine retribution awaits those who use such clever devices to subvert morality. It also becomes clear that Cooper, as he had in so many of his earlier books, employed his fiction to comment upon what he viewed as the political ills of the nation. He used the ending of *The Crater* to condemn what he believed were rampant abuses in the United States at the time, in particular what Cooper saw as a perversion of Constitutional law. "By 1847," writes McWilliams,

> Cooper's national pessimism was literally nihilistic. Such an ending can only be viewed as the outgrowth of years of brooding about the decline of the republic ... he had given increasingly strident attention to possible national disaster stemming from demagogues who took advantage of the changeability of constitutional law … one feels that the volcano that was to explode was within James Fenimore Cooper.[37]

---

35 Huet, *The Culture of Disaster*, 2.

36 Huet, *The Culture of Disaster*, 2.

37 McWilliams, "The Crater and the Constitution," 632.

These rationales for the violent conclusion of Cooper's book, the political and the environmental, are not mutually exclusive. The same greed and lack of accountability that afforded political overreach was also at the heart of environmental waste and abuse, a form of environmental overreach. In each case, Cooper warns that those who engage in such wrongs will have to answer to a higher power. Whether he believed in it or not, for the aging Cooper, the threat of some horrid divine punishment was a useful literary tool with which to convey the seriousness of his concerns about the republic's future both politically and environmentally. Huet writes that "[t]he history of disasters is thus also the history of humans wresting from the heavens the source and reason of their misfortunes,"[38] but it would seem that Cooper (in a gesture we still see today in the media's coverage of a disaster) refuses to relinquish the idea of a divine will behind such calamities.

Thus Cooper's allegorical adventure tale ends with an off stage act of divine punishment, one that demonstrates how in his late career his environmental imagination had become firmly linked with what was essentially a form of typology. The result is an odd distancing of human action from environmental reaction. In this equation, humanity's failing is only indirectly linked to the larger disaster, one that does not demonstrate the human capability to destroy the environment but rather becomes—through religious superstition—somehow further proof of an all-powerful God's ability to punish those who ostensibly defy his will. The upshot of such thinking is environmentally disastrous in that it becomes a convenient way to deny human-exacerbated environmental degradation (or at the very least poor future planning). Presented with no witnesses to the disaster, we are, like Mark, left to puzzle over the horrible fate that has befallen those too-proud colonists who had attempted to transplant the evils of mankind to this new world. Yet even this frustratingly ambivalent end, the final, colony destroying earthquake demonstrates a Deleuizian aspect: by drowning the selfish and unworthy colonists the book ends with an arguably beneficial catastrophe that may serve as a caution to others who would place humanity in the position of a god and therefore lack the humility necessary to value the non-economic worth of the environment.

Conclusion

As we have become more and more attuned to environmental degradation, as we have come to further realize that the human mark on the earth is both lasting and poisonous, writers and filmmakers have become more emboldened to popularize the idea of humanity's self-inflicted demise. It is easy enough to see this trend at play in any number of contemporary texts. Indeed a cottage industry of environmental catastrophe narratives has sprung up from the carbon laden smog of our current environmental crisis: Margaret Atwood's *Oryx and Crake* (2004)

[38] Huet, *The Culture of Disaster*, 6.

and *The Year of the Flood* (2010); Cormac McCarthy's *The Road* (2007); Jose Saramago's *Blindness* (1999); Karen Tei Yamashita's *Tropic of Orange* (1997); and Craig Thompson's *Habibi* (2011). Even the recent resurgence of zombie films, some pointing to human causation from medical experiments gone wrong, might be interpreted as a cultural indicator of a pervasive environmental dread. These texts bring to the surface a monstrousness that seems to hide in the core of humanity's brutal and figuratively cannibalistic treatment of each other and the land.

At the heart of many of these narratives is the irony that we—like the colonists of Cooper's *The Crater*—stand to suffer by embracing the same colonial powers that made our species so successful in the first place. Though it is a cliché to argue that humanity's rise is due to our adaptability, such is rarely the case. Instead we uncompromisingly adapt the environment to ourselves, and in so doing we alter, often with cumulatively disastrous results, the very being of the land. Critics like Patrick D. Murphy have argued, and it is not a hard case to make, that these narratives of our communal demise are active attempts to change public perception and to force change in environmental policy. As Murphy contends, writers of apocalyptic and dystopian narratives "have not been crying wolf ... they have been trying to head human disasters off at the pass, and should be recognized for their contributions."[39] They are the town criers, warning us through the medium of popular culture at a time when governments and big business, especially in the United States, seem to ignore the problem with the same vigor with which they exacerbate it.

This concern with natural catastrophe as an environmental issue has its origins in the work of Paulding, Cole, Cooper, and others who recognized the cultural value of crafting environmental and sociopolitical allegories that demonstrate how humanity's relationship with the land might begin as a struggle against a forbidding landscape, reach a pastoral midpoint, and then fall into a mire of immorality and waste. Ultimately their work reflects the environmental anxiety that accompanied the beginning of America's industrial age. It warns that unless checked by a form of environmental and personal humility, humanity risks falling into a mire of greed and shortsighted destruction. As each of these works makes clear, this hazard is nowhere so pronounced as when people are placed within the tinder box of a resource rich environment, where our tendency is to burn quickly, brightly, then out.

---

39 Murphy, *Ecocritical Explorations in Literary and Cultural Studies*, 171–2.

Afterword

In 1850, the year before his death, James Fenimore Cooper was again associated with the publication of an influential work of environmental literature. The book in question was called *Rural Hours*, and it was written by his daughter, Susan Fenimore Cooper. This literary daybook chronicled her experience of the natural world of Cooperstown, New York, and reveals a writer of considerable talent. Like her father, Susan excelled in rendering nature into words while transcending mere description to engage in the complexities associated with the problems of human ownership of the land. While *Rural Hours* never approached what we would now consider bestseller status, it was influential enough that it served, as Lawrence Buell puts it, as "both a source and a rival of Thoreau's *Walden*."[1]

The book's relation to another environmental classic, *The Pioneers*, is equally worth noting. Both deal with the natural history of Cooperstown. Both are filled with beautiful natural observations. Both at times contradict their admiration for the natural world with praise for destructive forms of development. Still, these two books are as often worlds apart, and that distinction is most apparent in the way Susan so fully embraces a non-fictional, personal mode. James Fenimore Cooper, on the other hand, concerns himself with matters of an ambitiously national scale. He presents us with an assortment of characters allegorically standing in for broad social and political trends. Susan takes a more individualistic approach. Her book, as she makes abundantly clear in its delivery, is the account of one person, in one place, over one year. Susan's modest book helped usher in the kind of nature writing that would become the hallmark of the second half of the nineteenth century, on up to the present day. *Rural Hours* represented something of a passing of the environmental torch, not so much from father to daughter as from fiction to nonfiction, and from a national to regional scale.

As I have illustrated in this book, between the 1780s and the 1840s, U.S. literature developed an increasingly mature ecological consciousness, one that had as much to do with the natural world as it did with humanity's potential to harm the very land it needs to survive. With the emergence of Emersonian and Thoreauvian thinking, the grounded debate over practical and sustainable land use found in works like *The Pioneers* and *The Prairie* gave way to a form of egocentric nature writing that for all of its merits took a far more self-centered view. For example, I want to believe Thoreau when he states he wishes to "speak a word for Nature,"[2] but he seems incapable of doing so without also devoting pages and pages to himself.

In the second half of the nineteenth century the environmental question moved from one of community-centered land practices to one of how best to foster an individual, spiritual relationship with what remained of a conquered landscape.

---

1 Buell, "Foreword," *Susan Fenimore Cooper: New Essays on Rural Hours and Other Works*, vii.

2 Thoreau, "Walking," 260.

After all, the well traveled woods outside of Concord were a far cry from, say, Bartram's unspoiled southeastern wilds, Brockden Brown's nightmarish Pennsylvania woods, or Cooper's sparsely settled New York wilderness. These earlier texts ask: How do we best utilize and preserve our natural world while also strengthening our nation? Such questions are at times addressed in the best of Transcendentalist works, but for the most part, the drawing board of environmental fiction was abandoned for a more self-absorbed, Transcendentalist saunter. These far more metaphysical investigations conceive of the natural world as a form of Emersonian "NOT ME,"[3] and they decenter the subject of the environment, arguing that while it should be part of the human experience, it is nevertheless separate from that experience.

As I have argued, much of the environmentally tinted fiction of the early republic viewed nature as evidence of an ever-present and inescapable environmental contact zone between people and the land. Human impact had more to do with practical land use strategies than recreation or explorations of the self. Contrary to this trend, the advent of Transcendentalist-influenced nature writing often presented the author as a kind of solitary environmental prophet. It was an abandonment not only of the power of fiction to teach us sophisticated environmental lessons but also of the value of living in and working with the land.

For the most part, American environmental fiction went into a period of dormancy until the end of the nineteenth century, when Realism/Naturalism began to gain a foothold on the literary marketplace. Mark Twain, Charles Chesnutt, Stephen Crane, Sarah Orne Jewett, and others started to write about a more tangible natural world, one in which everyday people lived and died in relation to their use of the land and animals. Still, not until the early twentieth century would the reading public again see works of fiction that approached the overt environmental consciousness of James Fenimore Cooper's *The Pioneers*. Willa Cather's *O Pioneers!*, John Steinbeck's *The Grapes of Wrath*, Zora Neale Hurston's *Their Eyes Were Watching God*, and any number of books and stories by William Faulkner, whose *Light in August; Go Down, Moses; The Sound and the Fury; Absalom, Absalom!; If I Forget Thee, Jerusalem*; and *As I Lay Dying*, to name just a few, contribute to a fictionalized, if all too accurate, portrait of environmental degradation.

In fact, many of these books were deeply informed by real-world environmental crises. From the rampant deforestation of the reconstructed South to the 1928 Lake Okeechobee hurricane; from the push to plow up the Great Plains through mechanized farming to the resulting Dust Bowl of the "Dirty Thirties," American writers increasingly turned to fiction to make sense of an environment that seemed, like the God of Cooper's *The Crater*, to be punishing them as much for their sociopolitical failings as for their failure to act as responsible stewards of the land. As we face a future plagued by scores of major environmental crises, it may prove wise for our culture to take a look over our shoulder to the writings of those who so long ago warned against humanity's "Wasty Ways."[4]

---

3 Emerson, *Nature*, 36.

4 Cooper, *The Pioneers*, 248.

Bibliography

A Walk and Conversation Between a Fond Father and His Little Son,: As They Took a Walk through the Fields and Meadows, &c.: For the Use of Children. Norwich, CT: Trumbull, 1802. Early American Imprints, Series II. Shaw-Shoemaker (1801–1819). Shaw-Shoemaker Digital Edition. American Antiquarian Society. Web. July 11, 2012.

Abrams, Robert E. *Landscape and Ideology in American Renaissance Literature.* Cambridge: Cambridge University Press, 2004.

Adams, Charles H. "Reading Ecologically: Language and Play in Bartram's *Travels*." *Southern Quarterly* 32.4 (1994): 65–74.

Adkins, Kaye. "'Foundation-Stones': Natural History for Children in *St. Nicholas Magazine*." In *Wild Things: Children's Culture and Ecocriticism*, edited by Sidney I. Dobrin and Kenneth B. Kidd, 31–47. Detroit: Wayne State University Press, 2004.

The Affecting History of the Babes in the Wood. Hartford, CT: Babcock, 1796. Early American Imprints, Series I. Evans (1639–1800). Evans Digital Edition. American Antiquarian Society. Web. 11 July 2012.

Allen, Elizabeth K. *Open-Air Sketching: Nineteenth-Century American Landscape Drawings in the Albany Institute of History and Art.* Albany, NY: State University of New York Press, 1998.

American Primer, The. Medford, MA: Coverly for Cambridge, 1799. Early American Imprints, Series I. Evans (1639–1800). Evans Digital Edition. American Antiquarian Society. Web. July 12, 2012.

Anderson, Douglas. "Bartram's *Travels* and the Politics of Nature." *Early American Literature* 25 (1990): 3–17.

Arac, Jonathan. *The Emergence of American Literary Narrative, 1820–1860.* Cambridge, MA: Harvard University Press, 2005.

Barnhill, Georgia B. "Transformations in Pictorial Printing." In *A History of the Book in*

America. Volume 2: An Extensive Republic: Print, Culture, and Society in the New Nation, 1790–1840. Edited by Robert A. Gross and Mary Kelley, 422–40. Chapel Hill, NC: University of North Carolina Press, 2010.

Bartram, William. *Botanical and Zoological Drawings, 1756–1788.* Edited by Joseph Ewan. Philadelphia: American Philosophical Society, 1968.

———. *Travels and Other Writings.* Edited by Thomas P. Slaughter. New York: Library of America, 1996.

Bauer, Ralph. *The Cultural Geography of Colonial American Literatures: Empire, Travel, and Modernity.* New York: Cambridge University Press, 2003.

Bellin, Joshua David. "Wicked Instruments: William Bartram and the Dispossession of the Southern Indians." *Arizona Quarterly* 52.3 (1995): 1–23.

Bentham, Jeremy. *An Introduction to the Principles of Morals and Legislation.* Oxford University Press, 1907.

Berthold, Dennis. "Charles Brockden Brown, *Edgar Huntly*, and the Origins of the American Picturesque." *William and Mary Quarterly* 41 (1984): 62–84.

———. "Desacralizing the American Gothic: An Iconographic Approach to *Edgar Huntly*." *Studies in American Fiction* 14 (1986): 127–38.

Bewell, Alan. "Jefferson's Thermometer: Colonial Biogeographical Constructions of the Climate of America." In *Romantic Science: The Literary Forms of Natural* History, edited by Noah Heringman, 111–38. Albany: State University of New York Press, 2003.

Bingham, Caleb. *The Hunters, or The Sufferings of Hugh and Francis, in the Wilderness: A True Story*. Boston: Armstrong for Bingham, 1814. Early American Imprints, Series II. Shaw-Shoemaker (1801–1819). Shaw-Shoemaker Digital Edition. American Antiquarian Society. Web. July 11, 2012.

Blanchot, Maurice. *The Writing of the Disaster*. Translated by Ann Smock. Lincoln: University of Nebraska Press, 1995.

Bleecker, Ann Eliza. *The History of Maria Kittle*. In *Women's Early American Historical Narratives*. Edited by Sharon M. Harris, 3–35. New York: Penguin, 2003.

Bradford, William. *Of Plymouth Plantation: 1620–1647*. Edited by Samuel Eliot Morrison. New York: Random House, 1952.

Branch, Michael P. "Before Nature Writing: Discourses of Colonial American Natural History." In *Beyond Nature Writing: Expanding the Boundaries of Ecocriticism*. Edited by Karla Armbruster and Katherine R. Wallace, 91–107. Charlottesville, VA: University of Virginia Press, 2001.

Brown, Charles Brockden. *Edgar Huntly, or Memoirs of a Sleep-walker*. Edited by Sydney J. Krause and S.W. Reid. Kent, OH: Kent State University Press, 1984.

Buel, Jesse. *The Farmer's Companion; or, Essays on the Principles and Practice of Animal Husbandry*. Boston: Marsh, Capen, Lyon, and Webb, 1839.

Buell, Lawrence. *The Environmental Imagination: Thoreau, Nature Writing, and the Formation of American Culture*. Cambridge, MA: Harvard University Press, 1995.

———. "Foreword." In *Susan Fenimore Cooper: New Essays on Rural Hours and Other Works*. Edited by Rochelle Johnson and Daniel Patterson, vii–viii. Athens: University of Georgia Press, 2001.

———. The Future of Environmental Criticism. Malden, MA: Blackwell, 2005.

Buffon, [Georges-Louis Leclerc,] Count de. Translated by William Smellie. 3rd ed. 9 vols. London: A. Strahan and T. Cadell, 1791.

Campbell, Alison, Valerie Kapos, and Jörn P.W. Scharlemann. *Review of the Literature on the Links Between Biodiversity and Climate Change: Impacts, Adaptation, and Mitigation*. Secretariat of the Convention of Biological Diversity. CDB Technical Series No. 42. Cambridge: United Nations Environment Programme, 2009.

Cañizares-Esguerra, Jorge. *How to Write the History of the New World: Histories, Epistemologies, and Identities in the Eighteenth-Century Atlantic World*. Palo Alto, CA: Stanford University Press, 2001.

Castiglia, Christopher. *Bound and Determined: Captivity, Culture-Crossing, and White Womanhood, from Mary Rowlandson to Patty Hearst*. Chicago: University of Chicago Press, 1996.

Clark, Harry Hayden. "Fenimore Cooper and Science." *Transactions of the Wisconsin Academy of Sciences, Arts, and Letters* 47 (1959): 179–204.

Cock Robin's Death and Funeral. Boston: Bible and Heart, 1780. Early American Imprints, Series I. Evans (1639–1800). Evans Digital Edition. American Antiquarian Society. Web. July 18, 2012.

Cocteau, Jean. *Professional Secrets: An Autobiography of Jean Cocteau*. Edited by Robert Phelps. New York: Farrar, Straus, and Giroux, 1970.

Cole, Thomas. "Essay on American Scenery." *American Monthly Magazine* 1 (1836): 1–12.

Cooper, James Fenimore. *The Crater; or, Vulcan's Peak*. Edited by Thomas Philbrick. Cambridge, MA: Harvard University Press, 1962.

———. *Gleanings in Europe: Italy*. Historical introduction and explanatory notes by John Conron and Constance Ayers Denne; text established by Constance Ayers Denne. Albany: State University of New York Press, 1981.

———. *The Letters and Journals of James Fenimore Cooper*. 6 vols. Edited by James Franklin Beard. Cambridge, MA: Harvard University Press, 1960–1968.

———. *The Pioneers, or the Sources of the Susquehanna; A Descriptive Tale*. Historical introduction and explanatory notes by James Franklin Beard; text established by Lance Schachterle, and Kenneth M. Andersen, Jr. Albany: State University of New York Press, 1980.

———. *The Prairie: A Tale*. Edited with historical introduction by James P. Elliott. Albany: State University of New York Press, 1985.

Cooper, William. *A Guide in the Wilderness; or the History of the First Settlements in the Western Counties of New York with Useful Instructions to Future Settlers*. Reprint of Dublin 1810. Cooperstown, NY, 1936.

Crèvecoeur, J. Hector St. John de. *Letters from an American Farmer* and *Sketches of Eighteenth-Century America*. Edited by Albert E. Stone. New York: Penguin, 1986.

Cronon, William. *Changes in the Land: Indians, Colonists, and the Ecology of New England*. New York: Hill and Wang, 2003.

———. "The Trouble with Wilderness; or, Getting Back to the Wrong Nature." In *Uncommon Ground: Rethinking the Human Place in Nature*, edited by William Cronon, 69–90. New York: Norton, 1995.

"Cruelty to Animals." In *Juvenile Magazine*. Utica, NY: Western Sunday School Union, 1828. American Antiquarian Society Historical Periodicals Collection. Web. 11 July 2012.

Davidson, Cathy N. *Revolution and the Word: The Rise of the Novel in America*. Expanded Edition. New York: Oxford University Press, 2004.

Dean, Bradley, P. "Natural History, Romanticism, and Thoreau." In *American Wilderness: A New History*. Edited by Michael Lewis, 73–89. New York: Oxford University Press, 2007.

Dekker, George, and John P. McWilliams. "Introduction." *Fenimore Cooper: The Critical Heritage*. Edited by George Dekker and John P. McWilliams Boston: Routledge and Kegan, 1985.

Derounian-Stodola, Kathryn Zabelle, and James Arthur Levernier. *The Indian Captivity Narrative, 1550–1900*. New York: Twayne, 1993.

Dobrin, Sidney I., and Kenneth B. Kidd. *Wild Things: Children's Culture and Ecocriticism*. Detroit: Wayne State University Press, 2004.

———, "Introduction: Into the Wild." In *Wild Things: Children's Culture and Ecocriticism*, edited by Sidney I. Dobrin and Kenneth B. Kidd, 1–15. Detroit: Wayne State University Press, 2004.

Drinnon, Richard. *Facing West: The Metaphysics of Indian-Hating and Empire-Building*. Minneapolis, MN: University of Minnesota Press, 1997.

Elliott, James P. "Historical Introduction." In *The Prairie: A* Tale, edited with historical introduction by James P. Elliott, xv–xxxiii. Albany: State University of New York Press, 1985.

Emerson, Ralph Waldo. *The Journals and Miscellaneous Notebooks of Ralph Waldo Emerson, 1843–1847*. Cambridge, MA: Harvard University Press, 1982.

———. *Nature and Selected Essays*. Penguin, 1982.

Fellows, Otis E., and Stephen F. Milliken. *Buffon*. New York: Twayne, 1972.

Fessenden, Tracy. *Culture and Redemption: Religion, the Secular, and American Literature*. Princeton: Princeton University Press, 2007.

Filson, John. *The Discovery, Settlement and Present State of Kentuke: And an Essay towards the Topography, and Natural History of That Important Country: To Which Is Added, an Appendix, Containing, I. The Adventures of Col. Daniel Boon, One of the First Settlers, Comprehending Every Important Occurrence in the Political History of that Province. II. The Minutes of the Piankashaw Council, Held at Post St. Vincents, April 15, 1784. III. An Account of the Indian Nations inhabiting within the Limits of the Thirteen Original States, Their Manners and Customs and the Reflections on their Origin. IV. The Stages and Distances between Philadelphia and the Falls of the Ohio; from Pittsburg to Pensacola and Several Other Places.—The Whole Illustrated by a New and Accurate Map of Kentuke, and the Country Adjoining, Drawn from Actual Surveys.* Wilmington, DE, printed for James Adams, 1784.

Fletcher, Angus, *A New Theory for American Poetry: Democracy, the Environment, and the Future of Imagination*. Cambridge, MA: Harvard University Press, 2004.

Flint, Timothy. *Recollections of the Last Ten Years; Passed in Occasional Residences and Journeyings in the Valley of the Mississippi from Pittsburg and the Missouri to the Gulf of Mexico, and from Florida to the Spanish Frontier*. Boston: Cummings, Hilliard, and Co., 1826.

Franklin, Wayne. *James Fenimore Cooper: The Early Years*. New Haven: Yale University Press, 2007.

Golden, James. L. and Alan L. Golden. *Thomas Jefferson and the Rhetoric of Virtue*. New York: Rowman and Littlefield, 2002.

Goodrich, Samuel Griswold. *The Child's Botany*. Boston: S. G. Goodrich, 1828.

———. *The Man and the Snake*. Hartford, CT: Roberts & Burr for Goodrich, 1819. Early American Imprints, Series II. Shaw-Shoemaker (1801–1819). Shaw Shoemaker Digital Edition. American Antiquarian Society. Web. 11 July 2012.

———. *Peter Parley's Juvenile Tales*. 2nd ed. Boston: Carter and Hendee, 1830.

———. *Peter Parley's Story of Little Marion*. Boston: S. G. Goodrich, 1830.

———. *Recollections of a Lifetime, or Men and Things I Have Seen*. 2. Vols. New York: Miller, Orton, Mulligan, 1856.

———. *The Tales of Peter Parley, About America*. Boston: S.G. Goodrich, 1827.

———. *The Two Doves and the Owl*. Hartford, CT: Roberts & Burr for Goodrich, 1819. Early American Imprints, Series II. Shaw-Shoemaker (1801–1819). Shaw-Shoemaker Digital Edition. American Antiquarian Society. Web. July 11, 2012.

Hall, David D. *Cultures of Print: Essays in the History of the Book*. Boston: University of Massachusetts Press, 1996.

Hallock, Thomas. *From the Fallen Tree: Frontier Narratives, Environmental Politics, and the Roots of a National Pastoral, 1749–1826*. Chapel Hill, NC: University of North Carolina Press, 2003.

Hamilton, Geoff, and Brian Jones, eds. *Encyclopedia of the Environment in American Literature*. Jefferson, NC: McFarland, 2013.

Harris, Sharon M. "Ann Eliza Bleecker, 1752–1783." In *Women's Early American Historical Narratives*, edited by Sharon M. Harris, 1–2. New York: Penguin, 2003.

Hell, Julia. "Imperial Ruin Gazers, or Why did Scipio Weep?" In *Ruins of Modernity*, edited by Julia Hell and Andreas Schönle, 169–92. Durham, NC: Duke University Press, 2010.

The Hermit of the Forest, and the Wandering Infants: A Rural Fragment. Boston: Hall, 1789. Early American Imprints, Series I. Evans (1639–1800). Evans Digital Edition. American Antiquarian Society. Web. July 11, 2012.

Hines, Maude. "'He Made *Us* Very Much Like the Flowers:' Human/Nature in Nineteenth-Century Anglo-American Children's Literature." In *Wild Things: Children's Culture and Ecocriticism*, edited by Sidney I. Dobrin and Kenneth B. Kidd, 16–30. Detroit: Wayne State University Press, 2004.

Holbo, Christine. "Imagination, Commerce, and the Politics of Association in Crèvecoeur's *Letters from an American Farmer*." *Early American Literature* 32 (1997): 20–65.

Huet, Marie-Hélène. *The Culture of Disaster*. Chicago: University of Chicago Press, 2012.

Imlay, Gilbert. *The Emigrants*. Edited by W.M. Verhoeven and Amanda Gilroy. New York: Penguin, 1998.

Irmscher, Christoph. *The Poetics of Natural History: From John Bartram to William James*. New Brunswick, NJ: Rutgers University Press, 1999.

Jefferson, Thomas. "Bernard Lacépède." In *Letters of the Lewis and Clark Expedition with Related Documents, 1783–1854*, edited by Donald Jackson, 15–16. 2nd ed. Urbana: University of Illinois Press, 1978.

———. "Letter to Buffon, Oct. 1, 1787." In *Writings*, 909–10. New York: Library of America, 1984.

———. *Notes on the State of Virginia*. In *Writings*. New York: Library of America, 1984.

Johnson, Rochelle L. *Passions for Nature: Nineteenth-Century America's Aesthetics of Alienation*. Athens, GA: University of Georgia Press, 2009.

Juvenile Magazine. Utica, NY: Western Sunday School Union, 1828. American Antiquarian Society Historical Periodicals Collection. Web. July 11, 2012.

Juvenile Magazine, or Miscellaneous Repository of Useful Information. Vol. 1. Philadelphia, 1802. Early American Imprints, Series II. Shaw-Shoemaker (1801–1819). Shaw-Shoemaker Digital Edition. American Antiquarian Society. Web. July 11, 2012.

Kilcup, Karen L. *Fallen Forests: Emotion, Embodiment, and Ethics in American Women's Environmental Writing, 1781–1924*. Athens, GA: University of Georgia Press, 2013.

Kolodny, Annette. *The Land Before Her: Fantasy and Experience of the American Frontiers, 1630–1860*. Chapel Hill, NC: University of North Carolina Press, 1984.

———. *The Lay of the Land: Metaphor as Experience and History in American Life and Letters*. Chapel Hill, NC: University of North Carolina Press, 1975.

Krause, Sydney J. "Historical Essay." In Charles Brockden Brown, *Edgar Huntly, or Memoirs of a Sleep-walker*, edited by Sydney J. Krause and S. W. Reid, 295–400. Kent: Kent State University Press, 1984.

———. "Penn's Elm and *Edgar Huntly*: Dark 'Instruction to the Heart.'" *American Literature* 66 (1994): 463–84.

Kutchen, Larry. "The 'Vulgar Thread of the Canvas': Revolution and the Picturesque in Ann Eliza Bleecker, Crevecoeur, and Charles Brockden Brown." *Early American Literature* 36.3 (2001): 395–426.

Lawrence, D.H. *Studies in Classic American Literature*. Edited by Ezra Greenspan, Lindeth Vasey, and John Worthen. Cambridge: University of Cambridge Press, 2003.

Lerer, Seth. *Children's Literature: A Reader's History, From Aesop to Harry Potter*. Chicago: University of Chicago Press, 2008.

Lewis, Andrew J. "Gathering for the Republic: Botany in Early Republic America." In *Colonial Botany: Science, Commerce, and Politics in the Early Modern World*, edited by Londa Schiebinger and Claudia Swan, 66–80. Philadelphia: University of Pennsylvania Press, 2005.

Looby, Christopher. "The Constitution of Nature: Taxonomy as Politics in Jefferson, Peale, and Bartram." *Early American Literature* 22 (1987): 252–73.

Lorain, John. *Nature and Reason Harmonized in the Practice of Husbandry*. Philadelphia: Carey and Lea, 1825.

Love, Glen A. *Practical Ecocriticism: Literature, Biology, and the Environment*. Charlottesville, VA: University of Virginia Press, 2003.

Lowes, John Livingston. *The Road to Xanadu*. Princeton: Princeton University Press, 1986.

Lueck, Beth. "Charles Brockden Brown's *Edgar Huntly*: The Picturesque Traveler as Sleepwalker." *Studies in American Fiction* 15 (1987): 25–42.

MacDougall, Hugh. *The Cooper Bookshelf*. http://external.oneonta.edu/cooper/writings/journal.html#crater. Web. 2013.

Marsh, George Perkins. *Man and Nature; or, Physical Geography as Modified by Human Action*. New York: Scribner, 1864.

Marshall, Ian. *Story Line: Exploring the Literature of the Appalachian Trail*. Charlottesville, VA: University Press of Virginia, 1998.

Mazel, David. *American Literary Environmentalism*. Athens, GA: University of Georgia Press, 2000.

McWilliams, John P. "*The Crater* and the Constitution." *Texas Studies in Literature and Language* 12.4 (1971): 631–45.

Memorial of James Fenimore Cooper. New York: Putnam, 1852.

Menefee, Joan. "Good Masters: Child-Animal Relationships in the Writings of Mark Twain and G. Stanley Hall." In *Enterprising Youth: Social Values and Acculturation in Nineteenth-Century*, edited by Monika Elbert, 227–44. New York: Routledge, 2008.

Merchant, Carolyn. *American Environmental History: An Introduction*. New York: Columbia University Press, 2007.

———. Reinventing Eden: The Fate of Nature in Western Culture. 2nd ed. New York: Taylor and Francis, 2013.

Millhouse, Barbara Babcock. *American Wilderness: The Story of the Hudson River School of Painting*. Delmar, NY: Black Dome Press, 2007.

Monagham, Charles, and E. Jennifer Monaghan. "Schoolbooks." In *A History of the Book in America. Volume 2: And Extensive Republic: Print, Culture, and Society in the New Nation, 1790–1840*, edited by Robert A. Gross and Mary Kelley, 304–18. Chapel Hill, NC: University of North Carolina Press, 2010.

Moore, Bryan L. *Ecology and Literature: Ecocentric Personification from Antiquity to the Twenty-First Century*. New York: Palgrave, 2008.

Moore, Dennis D. Introduction. *More Letters from the American Farmer: An Edition of the Essays in English Left Unpublished by Crèvecoeur*. Athens, GA: University of Georgia Press, 1995.

Moore, L. Hugh. "The Aesthetic Theory of William Bartram." *Essays in the Arts and Sciences* 12.1 (1983): 17–35.

Morin, Peter J. *Community Ecology*. Malden, UK: Blackwell Science, 1999.
Morton, Timothy. *Ecology without Nature: Rethinking Environmental Aesthetics*. Cambridge, MA: Harvard University Press, 2007.
The Mountain Piper; or, The History of Edgar and Matilda. Worcester, MA: Isaiah Thomas, 1794. Early American Imprints, Series I. Evans (1639–1800). Evans Digital Edition. American Antiquarian Society. Web. July 11, 2012.
Murphy, Patrick D. *Ecocritical Explorations in Literary and Cultural Studies*. New York: Lexington, 2009.
Murphy, Thomas K. *A Land without Castles: The Changing Image of America in Europe, 1780–1830*. Lanham, MD: Lexington Books, 2001.
Nash, Roderick. *Wilderness and the American Mind*. 3rd ed. New Haven: Yale University Press, 1982.
New England Primer, The. Boston: Kneeland and Green, 1727. Early American Imprints, Series I. Evans (1639–1800). Evans Digital Edition. American Antiquarian Society. Web. July 12, 2012.
Noble, Louis Legrand. *The Life and Works of Thomas Cole*. Ed. Elliot S. Vesell. Delmar, NY: Black Dome Press, 1997.
"On a Goldfinch, Starved to Death in Its Cage." *Juvenile Magazine*. Utica, NY: Western Sunday School Union, 1828. American Antiquarian Society Historical Periodicals Collection. Web. July 11, 2012.
Opie, John. *Nature's Nation: An Environmental History of the United States*. New York: Harcourt, 1998.
Osborne, Jeff. "American Antipathy and the Cruelties of Citizenship in Crèvecoeur's *Letters from an American Farmer*." *Early American Literature* 42.3 (2007): 529–53.
Överland, Orm. *James Fenimore Cooper's The Prairie: The Making and Meaning of an American Classic*. Oslo: Universitetsforlaget, 1973.
Panther, Abraham [pseud.]. *A Surprising Account of the Discovery of a Lady Who Was Taken by the Indians in 1777, and After Making Her Escape, She Retired to a Lonely Cave, Where She Lived Nine Years. In Women's Indian Captivity Narratives*. Edited by Kathryn Zabelle Derounian-Stodola, 86–90. New York: Penguin, 1998.
Parley's Magazine. Vol. 1. Boston: Lilly, Wait and Co.; Portland, ME: Colman, Holden, and Co. 1833. Early American Imprints, Series II. Shaw-Shoemaker (1801–1819). Shaw-Shoemaker Digital Edition. American Antiquarian Society. Web. July 13, 2012.
Paulding, James Kirke. *The Backwoodsman. A Poem*. Thomas: Philadelphia, 1818.
———. *Westward Ho!*. New York: Harper, 1832.
Penrick, James L. *The New Madrid Earthquakes*. Columbia: University of Missouri Press, 1981.
Perrin, Anne. "Opened Frontiers, Closed Deserts: The Contradictions between Source and Text in James Fenimore Cooper's *The Prairie*." In *James Fenimore Cooper: His Country and His Art, Papers from the 2001 Cooper Seminar*, edited by Hugh C. MacDougall, 72–76. Oneonta, NY, 2001.

Pictures of Seventy-Two Beasts and Birds, with a Familiar Description of Each, In Verse and Prose, Published for the Benefit of the Lovers of Natural History, and for the Instruction of the Youth of America. Boston: Samuel Hall, 1796. Early American Imprints, Series I. Evans (1639–1800). Evans Digital Edition. American Antiquarian Society. Web. July 16, 2012.

Pitcher, E.W. "The 'Hapless Babes' of the Frontier: Ovid, 'The History of Maria Kittle,' and *The Last of the Mohicans*." *ANQ* 13.3 (2000): 33. *Literature Resource Center*. Web. September 24, 2013.

Pratt, Amy. "The Pleasure of Being Lost: 'The Panther Captivity' and the Metaphysics of Commerce." *Mosaic* 34.1 (2001). Literature Resource Center. Web. September 23, 2013.

Regis, Pamela. *Describing Early America: Bartram, Jefferson, Crèvecoeur, and the Rhetoric of Natural History*. DeKalb: Northern Illinois University Press, 1992.

Reice, Seth R. *The Silver Lining: The Benefits of Natural Disaster*. Princeton, NJ: Princeton University Press, 2001.

Rice, Grantland S. "Crèvecoeur and the Politics of Authorship in Republican America." *Early American Literature* 28.2 (1993): 91–119.

Ringe, Donald. *James Fenimore Cooper*. Boston: Twayne, 1962.

Robinson, E. Arthur. "Conservation in Cooper's *The Pioneers*." *PMLA* 82.7 (1967): 564–78.

Roger, Jacques. *Buffon: A Life in Natural History*. Ithaca, NY: Cornell University Press, 1997.

Roger, Philippe. *The American Enemy: The History of French Anti-Americanism*. Translated by Sharon Bowman. Chicago: University of Chicago Press, 2005.

Roselle, Daniel. *Samuel Griswold Goodrich, Creator of Peter Parley: A Study of His Life and Work*. Albany, NY: State University of New York Press, 1968.

Rosenbach, A.S.W. *Early American Children's Books*. Portland, ME: Southworth Press, 1933.

Rutkow, Eric. *American Canopy: Trees, Forests, and the Making of a Nation*. New York: Scribner, 2012.

Sachs, Aaron. *Arcadian America: The Death and Life of an Environmental Tradition*. New Haven: Yale University Press, 2013.

Savoy, Eric. "The Face of the Tenant: A Theory of American Gothic." In *American Gothic: New Interventions in a National Narrative*, edited by R. K. Martin and E. Savoy, 3–19. Iowa City: University of Iowa Press, 1998.

Schachterle, Lance. "On *The Prairie*." In *Leather-Stocking Redux; or, Old Tales, New Essays*, edited by Jeffrey Walker, 124–49. New York: AMS Press, 2009.

Schoolcraft, Henry R. *Transallegania; or, the Groans of Missouri*. New York: Seymour, 1820.

Scudder, Thayer. *Global Threats, Global Futures: Living with Declining Living Standards*. Northhampton, MA: Elgar, 2010.

Seelye, John. *Beautiful Machine: Rivers and the Republican Plan, 1755–1825*. New York: Oxford University Press, 1991.

———. "The Jacobin Mode in Early American Fiction: Gilbert Imlay's *The Emigrants*." *Early American Literature*: 22.2 (1987): 204–12.
Shields, David S. "The Learned World." In *A History of the Book in America. Volume 2: An Extensive Republic: Print, Culture, and Society in the New Nation, 1790–1840*, edited by Robert A. Gross and Mary Kelley, 247–65. Chapel Hill, NC: University of North Carolina Press, 2010.
Silver, Bruce. "William Bartram's and Other Eighteenth-Century Accounts of Nature." *Journal of the History of Ideas* 39 (1978): 597–614.
Sivils, Matthew Wynn. "'The Base, Cursed Thing': Panther Attacks and Ecotones in Antebellum American Fiction." *Journal of Ecocriticism* 2.1 (2010): 19–32.
Slaughter, Thomas P. *The Natures of John and William Bartram*. New York: Knopf, 1996.
Smith, Henry Nash. *Virgin Land: The American West as Symbol and Myth*. Cambridge, MA: Harvard University Press, 1978.
Smith, Sydney. *The Works of the Rev. Sydney Smith*. New York: Appleton, 1870.
Smith, Thomas Ruys. *River of Dreams: Imagining the Mississippi Before Mark Twain*. Baton Rouge: Louisiana State University Press, 2007.
Spradlin, Derrick. "Westward Expansion, Indian Subjugation, and the Frontier Gazes of Jackson Johonnet, Zebulon Montgomery Pike, and James Kirke Paulding." *Essays in Arts and Sciences* 34.1 (2005): 39–50.
Stahl, J. D. "Natural History for Children and the Agassiz Association." In *Enterprising Youth: Social Values and Acculturation in Nineteenth-Century*, edited by Monika Elbert, 213–26. New York: Routledge, 2008.
Stoll, Steven. "Farm Against Forest." In *American Wilderness: A New History*, edited by Michael Lewis, 55–72. New York: Oxford University Press, 2007.
Stone, Albert E. Introduction. *Letters from an American Farmer* and *Sketches of Eighteenth-Century America*. Edited by Albert E. Stone. New York: Penguin, 1986.
Sweet, Timothy. *American Georgics: Economy and Environment in Early American Literature*. Philadelphia: University of Pennsylvania Press, 2002.
Taylor, Alan. *William Cooper's Town: Power and Persuasion on the Frontier of the Early American Republic*. New York: Knopf, 1995.
Thoreau, Henry David. *The Journal, 1837–1861*. Ed. Damion Searls. New York: New York Review of Books, 2009.
———. "Walking." In *Walden, Civil Disobedience, and Other Writings*. Edited by William Rossi. 260–87. New York: Norton, 2008.
Trimmer, Sarah. *Fabulous Histories, Designed for the Amusement & Instruction of Young Persons*. Philadelphia: Gibbons, 1794. Early American Imprints, Series I. Evans (1639–1800). Evans Digital Edition. American Antiquarian Society. Web. July 18, 2012.
Trumbull, Louisa Jane. *Worcester Through a Child's Eyes: The Diaries of Louisa Jane Trumbull, 1829–1837*. Ed. Holly V. Izard. Worcester, MA: American Antiquarian Society, 2005.
Tuan, Yi-Fu. *Escapism*. Baltimore, MD: Johns Hopkins University Press, 1998.

Uhl, Christopher. Developing Ecological Consciousness: The End of Separation. 2ne ed. New York: Rowman, 2013.

VanDerBeets, Richard. *The Indian Captivity Narrative: An American Genre*. Lanham, MD: University Press of America, 1984.

Verhoeven, W.M., and Amanda Gilroy. "Introduction." In *The Emigrants* by Gilbert Imlay, edited by W. M. Verhoeven and Amanda Gilroy, ix–xlix. New York: Penguin, 1998.

Verhoeven, Wil. "'Some corner of a foreign field that is forever England': The Transatlantic Construction of the Anglo-American Landscape." *Green and Pleasant Land: English Culture and the Romantic Countryside*. Edited by Amanda Gilroy, 155–71. Bondgenotenlaan, Belgium: Peeters, 2004. Waselkov, Gregory A., and Kathryn E. Holland Braund. *William Bartram on the Southeastern Indians*. Lincoln: University of Nebraska Press, 1995.

Webster, Noah. *History of Animals; being the Fourth Volume of Elements of Useful Knowledge. For the use of schools, and young persons of both sexes*. New Haven, CT: Walter and Steele, 1812. Early American Imprints, Series II. Shaw-Shoemaker (1801–1819). Shaw-Shoemaker Digital Edition. American Antiquarian Society. Web. 11 July 2012.

Welch, d'Alté A. *A Bibliography of American Children's Books Printed Prior to 1821*. Worcester, MA: American Antiquarian Society, 1972.

West, Rinda. *Out of the Shadow: Ecopsychology, Story, and Encounters with the Land*. Charlottesville, VA: University of Virginia Press, 2007.

White, Ed. "Crèvecoeur in Wyoming." *Early American Literature*: 43.2 (2008): 379–407.

White, P.S., and S.T.A. Pickett. "Natural Disturbance and Patch Dynamics: An Introduction." In *The Ecology of Natural Disturbance and Patch Dynamics*, edited by S. T. A. Pickett and P. S. White, 3–13. Orlando: Academic Press, 1985.

Williams, James. "Deleuze on J. M. W. Turner: Catastrophism in Philosophy." In *Deleuze and Philosophy: The Difference Engineer*, edited by Keith Ansell-Pearson, 233–.246. New York: Routledge, 1997.

Willis, Lloyd. *Environmental Evasion: The Literary, Critical, and Cultural Politics of "Nature's Nation."* Albany, NY: State University of New York Press, 2011.

"The Wolf." *Juvenile Magazine*. Utica, NY: Western Sunday School Union, 1828. American Antiquarian Society Historical Periodicals Collection. Web. 11 July 2012.

Women's Indian Captivity Narratives. Edited by Kathryn Zabelle Derounian-Stodola New York: Penguin, 1998.

Worster, Donald. *Nature's Economy: A History of Ecological Ideas*. 2nd ed. Cambridge: Cambridge University Press, 1994.

Ziser, Michael. *Environmental Practice and Early American Literature*. Cambridge: Cambridge University Press, 2013.

Index

FSC
www.fsc.org
MIX
Papier | Fördert gute Waldnutzung
FSC® C083411

Kolibri 360 GmbH
Ferdinand-Jühlke-Straße 7
99095 Erfurt, Deutschland
produktsicherheit@kolibri360.de